FUNDAMENTALS OF HORTICULTURE

THE AUTHORS

Mr. Ravi Partap Singh was born 31 Oct. 1994 at Kilaie, village, in Pratapgarh district of Uttar Pradesh. He did his B.Sc. (Ag.) from Udai Pratap Autonomous college Varanasi (U.P.) in 2012-16. He completed his M.Sc. (Ag.) Horticulture from department of Horticulture and forestry, ANDUAT Kumarganj, Ayodhya in 2016-2018. At present, he is pursuing Ph.D. Horticulture from College of Horticulture and forestry, ANDUAT Kumarganj, Ayodhya. He has also qualified ASRB-NET (National Eligibility Test) in Fruit Science (2018). Author has 7 Research papers, 1 Book, 8 book chapters, 15 popular articles go to his credit, which have been published.

Dr. Arun Kumar Singh was born in 1959 at Varanasi, Uttar Pradesh, received his B.Sc. (Ag.) in 1980, M.Sc. (Ag.) Horticulture in 1982 and earned his Ph.D. in Horticulture, from Institute of Agriculture Science, Banaras Hindu University. Dr. Singh started his career as an Assistant Professor, Horticulture at A.N.D. University of Agriculture and Technology, Kumarganj, Ayodhya, U.P in 1987 and become an Associate Professor in 1998. Presently he works as Horticulture and Landscape Officer and the Head, Floriculture and Landscape. He has supervised and mentored more than 35 M.Sc. and 12 Ph.D. students in his academic career. Dr. Singh has published more than 70 research papers of National and International journals, 2 book, 2 book chapters, 1 teaching manual, 40 popular articles.

Mr. Anshuman Singh was born 20 July, 1995 at Samodhpur, Village, in Jaunpur district of Uttar Pradesh. He did his B.Sc. (Hons.) Horticulture from BUAT Banda (U.P.) in 2012-16. He completed his M.Sc. (Ag.) Horticulture from department of Horticulture and forestry, ANDUAT Kumarganj, Ayodhya in 2016-2018. At present, he is pursuing Ph.D. Horticulture from department of Horticulture and forestry, ANDUAT Kumarganj, Ayodhya. He has also qualified ASRB-NET (National Eligibility Test) in Fruit Science (2018).

FUNDAMENTALS OF HORTICULTURE

Practical Manual

Ravi Pratap Singh*
Ph.D. Scholar Horticulture,
A.N.D.U.A & T, Ayodhya (U.P.)

Arun Kumar Singh
Head Department of Floriculture
and Landscape,
A.N.D.U.A&T, Ayodhya (U.P)

Anshuman Singh
Ph.D. Scholar Horticulture,
A.N.D.U.A&T, Ayodhya (U.P.)

2022

Daya Publishing House®
A Division of
Astral International Pvt. Ltd.
New Delhi – 110 002

ISBN: 9789354614750

Published by : **Daya Publishing House®**
A Division of
Astral International Pvt. Ltd.
– ISO 9001:2015 Certified Company –
4736/23, Ansari Road, Darya Ganj,
New Delhi-110 002
Ph. 011-43549197, 23278134
E-mail: info@astralint.com
Website: www.astralint.com

Preface

Horticulture, covering a gamut of crops ranging from herbaceous annual to woody perennials, constitutes a viable discipline in the area of Agricultural sciences. A series of practical courses in the subject are highly ramified and need specific literature for a particular matter. The text narrating introductory things are largely scattered. The students have to turn-out the pages of numerous practical manual to cover the matter on diverse practical subject of Horticulture. It has been in this context that work of compiling various facts of horticulture has been attempted and I feel pleased to present it in the form of the practical manual to the hands of the readers. Encompassing twelve(12) chapters, the practical manual is a detailed and comprehensive summery of the facts and figours yet known. However, the things are not only limited to the ecosphere that are known to mankind. Even beyond that they exist into cosmos also which are hard to observe and glance through the simple eyes. There may be the chances of lacks and lacuna. Thus scope for further improvement always rests. I shall be highly obliged to all such readers who will be extending their fellow-felling in further improvement of the present endeavour. The author express his gratefulness to all those whose deeds have been useful directly or indirectly in this compilation. I feel indebted to my Advisor Prof. Arun Kumar Singh, HOD, floriculture and landscaping, Kumarganj and Prof. Dharmendra Kumar Singh, HOD, Horticulture, Udai Pratap Autonomous College, Varanasi who ushered me the tottering talk of the subject and it is due to their blessings that I have been able to get the manuscript compiled. I don't find words sufficient to express my indebtedness to my parents, brothers and sisters, whose blessings and inspirations enforced me to complete this project. It is with due reverence that I wish to acknowledge Dr. O.P. Rao, Dean, COH, Kumarganj, whose work embracing chastity always provided an impetus to me. I have great pleasure in acknowledging moral support extended by our esteemed professors, Prof. Sanjay Pathak, HOD, fruit science, ANDUAT, Kumarganj, and Principal Scientist Kanchan Kumar Shrivastava, CISH, Lucknow and all the teachers of the College of Horticulture, Kumarganj, Ayodhya. I am very much confident that this Practical Manual "Fundamentals of Horticulture" would find a place of preference among various categories of readers in Horticulture.

RAVI PRATAP SINGH

Contents

Contents

Exercise No. 1 **Date : ____________**

OBJECTIVE: IDENTIFICATION OF GARDEN TOOLS

Although many tools and implements may be used for a single horticulture operation, Use of a right type of tool for a particular operation to achieve maximum efficiency and quick turnout of work is indispeusible. Selection of suitable tools/ implements is necessary for carrying out various horticultural operations. For example several tools are there for cutting branches of plants such as secature, tree pruner, pruning saw, carpenters saw etc. If the thickness of branch is about 1-2 cm, the secature may be more suitable, similarly if the branch in 4-6 cm is thickness pruning saw is the proper tool and if the branch is still bigger (more than 8cm) carpenters saw may be required. Therefore one has to exercise option to select a suitable tool to carry out various operations from the stage of land preparation to harvesting.

The following tools, and plant protection equipments are commonly used in the horticultural crops.

Sl.No.	Name	Diagrams
1.	**Bill hook** It is a hook shaped tool having single or double cutting edge, consisting of a curved blade attached with a plastic or wooden handle. It has a blade made of high carbon steel. The blade's length is 13 cm and width 2 cm. Bill hook is used for cutting shrubs and lopping old or dead branches of a tree. It is used for carrying out heavy pruning operations.	
2.	**Budding-cum-grafting knife** A budding cum grafting knife is a combination of two knives used for carrying out budding and grafting operations. It consists of two blades, one for carrying out budding and the other for grafting. These knives are fixed to the ends of a handle. Both the knives are made of high carbon steel. The knives are foldable into the handle. The length of the knife may be 6.5–7.5 cm and width 1.5 cm.	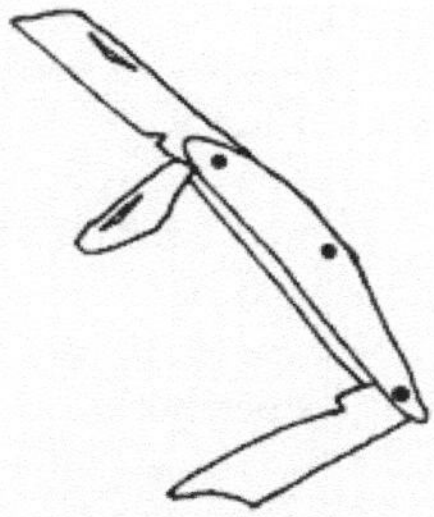

3.	**Flower scissors** Flower scissors or cutters are used for cutting flowers along with the stems. A flower scissor has two short blades with handles.	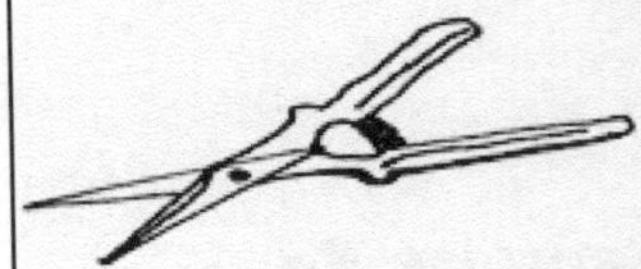
4.	**Rake or garden rake** It is used for breaking up the soil surface into a fine tilth, ready for sowing and collecting weeds and stones.	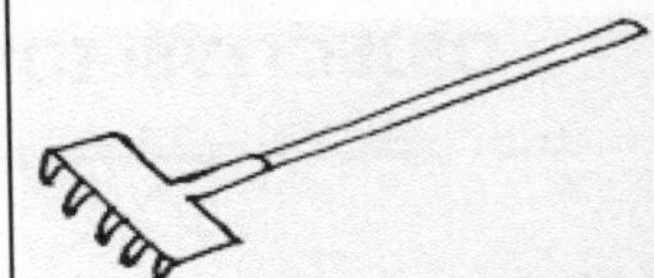
5.	**Grass shear** Grass shear is used to maintain a lawn. It is used for trimming and side dressing of the lawn. The most important part of a grass shear are the cutting blades, which are made of high carbon or alloy steel. The blades are sharp at the cutting edges. These are joined to a 'V' shaped spring steel handle, which always keeps the shearing blades open. Cutting takes place due to the shearing action of the blades. The length of the blade varies and is, usually, 15 20 cm.	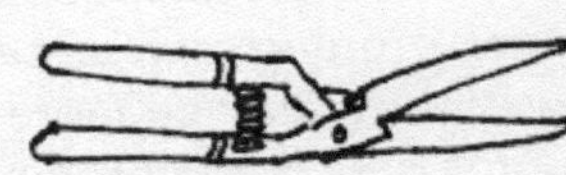
6.	**Hand cultivator** Hand cultivator It is used for altering and loosening the soil without causing damage to the roots of plants in a garden or nursery.	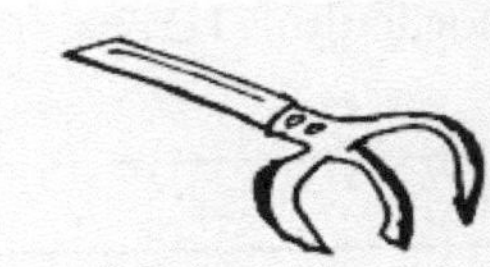
7.	**Hedge shear** It is used for performing various garden operations like pruning, cutting and trimming of hedge and shrubs, and making them look attractive. It consists of two cutting blades with tongs. These are made of high carbon steel in a single piece. The tongs are inserted for better grip in a wooden handle. The size of the shear is according to the size of the blades, varying from 15 to 30 cm in length and 0.8 cm in thickness.	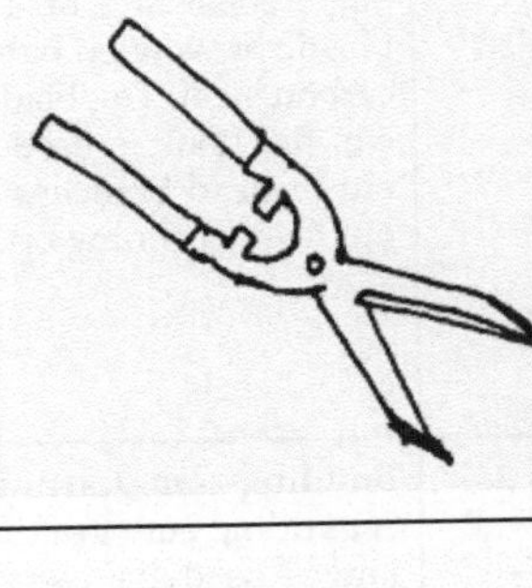
8.	**Khurpi** It is made of cast iron with a wooden handle attached to one side. It is meant for weeding, lifting of seedlings and plants in a nursery, transplanting plants in pots and field, and performing various other gardening operations.	

9.	**Lawn mower** To cut the grass uniformly in the lawn. It is having a roller behind to pad the grass to have cushion.	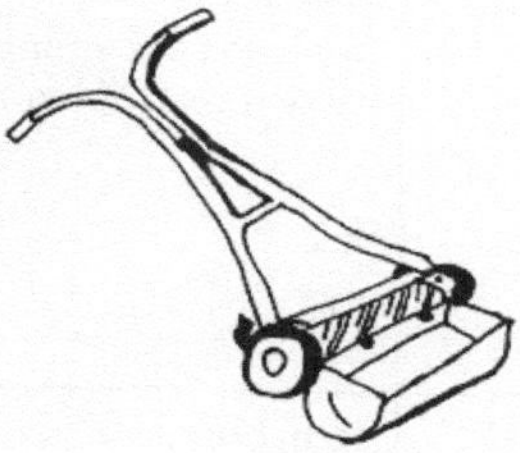
10.	**Pruning saw** It is used to cut crotch (angle) thicker branches of plants.	
11.	**Pruning or slashing knife** It is mostly used for removing unwanted and dense branches or twigs on plants. The knife made of carbon or alloy steel is tightly fit into the wooden or plastic handle. The tip of the knife is slightly hooked or curved for easy cutting or slashing small branches.	
12.	**Secateur** Secateur is meant for cutting branches, de shooting, disbudding, cutting of scion sticks, defoliation of leaves from the sticks, topping off small trees, etc. It is also useful in pruning pencil thick branches and making cuttings for propagation.	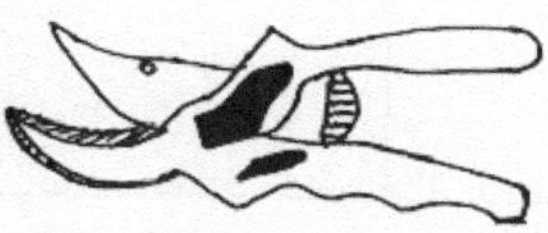
13.	**Sprayer** It is used for spraying insecticides, fungicides, herbicides, fertilisers and various other chemicals in a field. A variety of sprayers are available in the market to suit the requirements of different plants.	
14.	**Spade** It is made of cast iron and is used for digging or turning over the soil, making bunds in the field and small plots, etc.,	
15.	**Transplanting trowel** To lift the young seedlings along with a ball of earth for transplanting.	

16 .	**Tree pruner** It is used for pruning the shoots of trees, which are beyond reach from the ground level	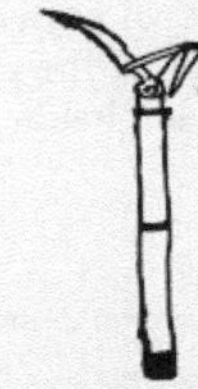
17.	**Trenching hoe** It is a small handle with a paddle and blade at the end. It is useful for cultivating the garden soil and carrying out weeding operations. There are different types of garden hoe made for specific uses	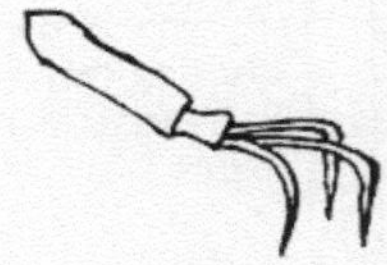
18.	**Water can** It is used for watering seedbeds, nursery beds and potted plants to avoid washing off the soil and causing damage to young seedlings.	
19.	**Wheel barrow** It can be used to move heavy things from one place to another, and also to collect trash in the garden and to transport manures, soil, seedlings, garden waste etc.	

Study questions:

1. Why tools and implements are necessary for gardens?
2. Draw a neat diagrames of different tools / implements / equipments studied in class
3. List out the tools/implements/equipments used for:
   (a) Digging
   (b) Planting and transplanting
   (c) Inter-cultivation or Interculture
   (d) Training and pruning
   (e) Irrigation
   (f) Weeding
   (g) Plant protection purpose.

Exercise No. 2 **Date : ____________**

IDENTIFICATION OF HORTICULTURAL CROPS

A. FRUIT CROPS

S.No.	English Name	Botanical Name	Family
1.	Aonla	*Phyllanthus emblica Gaerth*	Euphorbiaceae
2.	Almond	*Prunus dulcis Mill.*	Rosaceae
3.	Apple	*Malus pumila Mill.*	Rosaceae
4.	Apricot	*Prunus armeniaca L.*	Rosaceae
5.	Atemoya	*Annona atemoya Hort.*	Annonaceae
6.	Avocado	*Persea americana Mill.*	Lauraceae
7.	Bael	*Aegle marmelos Correa.*	Rutaceae
8.	Banana	*(edible fruit) Musa sapientum L.*	Musaceae
9.	Banana	*(Plantain cooking) Musa paradisiaca L.*	Musaceae
10.	Bardados cherry	*Malpighia glabra L.*	Malpighiaceae
11.	Black berry	*Rubus canadensis L.*	Rosaceae
12.	Carambola	*Averrhoa carambola L.*	Oxalidaceae
13.	Cherimoya	*Annona cherimola Mill.*	Annonaceae
14.	Cherry (Sweet)	*Prunus avium L.*	Rosaceae
15.	Chestnut	*Castanea sativa Mill.*	Fagaceae
16 .	Chinese Ber	*Zizyphus jujube Lamk.*	Rhamnaceae
17.	Custard apple	*Annona squamosa L.*	Annonaceae
18.	Date	*Phoenix dactylifera L.*	Palmaceae
19.	Durian Civet fruit	*Durio zibethinus L.*	Bombacaceae
20.	Fig	*Ficus carica L.*	Moraceae
21.	Filbert nut	*Corylus avellana Mill.*	Betulaceae
22.	Kiwi fruit	*Actinidia chinensis*	Dilleniaceae
23.	Grape (European)	*Vitis vinifera L.*	Vitaceae
24.	Guava	*Psidium guajava L.*	Myrtaceae
25.	Hazel nut	*Corylus avellana Mill.*	Betulaceae
26.	Jackfruit	*Artocarpus heterophyllus Lam.*	Moraceae
27.	Jamun	*Syzygium cuminii Skeels*	Myrtaceae
28.	Karonda	*Carissa carandas L.*	Apocynaceae

29.	Lasoda	*Cordia myxa Roxb.*	Boraginaceae
30.	Litchi	*Litchi chinensis Sonn.*	Sapindaceae
31.	Loquat	*Eriobotrya japonica Lindl.*	Rosaceae
32.	Macadamia nut	*Macadamia ternifolia Muell.*	Protenaceae
33.	Mandarin	*Citrus reticulata Blanco.*	Rutaceae
34.	Mango	*Mangifera indica L.*	Anacardiaceae
35.	Mangosteen	*Garcinia mangostana L.*	Guttiferae
36.	Mulberry (Black)	*Morus nigra L.*	Moraceae
37.	Mulberry (White/Green)	*Morus alba L.*	Moraceae
38.	Olive	*Olea europaea L.*	Oleaceae
39.	Papaya	*Carica papaya L.*	Caricaceae
40.	Passion fruit	*Passiflora edulis Sims.*	Passifloraceae
41.	Peach	*Prunus persica L.*	Rosaceae
42.	Pear (European)	*Pyrus communis L.*	Rosaceae
43.	Pecan nut	*Carya illinoinensis Koch.*	Juglandaceae
44.	Persimmon	*Diospyros kaki L.*	Ebenaceae
45.	Phalsa	*Grewia subinequalis*	Tiliaceae
46.	Pineapple	*Ananas comosus Merr.*	Bromeliaceae
47.	Pistachio nut	*Pistacia vera L.*	Anacardiaceae
48.	Plum (European)	*Prunus domestica L.*	Rosaceae
49.	Plum (Japanese)	*Prunus salicina*	Rosaceae
50.	Pomegranate	*Punica granatum L.*	Punicaceae
51.	Grapefruit	*Citrus paradisi Macf.*	Rutaceae
52.	Pummelo	*Citrus maxima Osbeck*	Rutaceae
53.	Rambutan	*Naphelium lappaceum*	Sapindaceae
54.	Sapota	*Achras zapota L.*	Sapotaceae
55.	Strawberry	*Fragaria chiloensis Duch.*	Rosaceae
56.	Sweet orange	*Citrus sinensis L.*	Rutaceae
57.	Tamarind	*Tamarindus indica L.*	Leguminosae
58.	Walnut	*Juglans regia L.*	Juglandaceae.
59.	Wood apple	*Feronia limonia L.*	Rutaceae

B. VEGETABLE CROPS

S.No.	English Name	Botanical Name	Family
Class Monocotyledoneae			
1.			
2.	Sweet corn	*Zea mays var. rugosa*	Gramineae
3.	Onion	*Allium cepa*	Alliaceae
4.	Leek	*Allium porrum*	Alliaceae

5.	Garlic	*Allium sativum*	Alliaceae
6.	Asparagus	*Asparagus officinalis*	Liliaceae
7.	Arvi or taro	*Colocasia esculenta*	Araceae
8.	Lesser yam	*Dioscorea esculenta*	Dioscoreaceae
9.	Greater yam	*Dioscorea rotundata*	Dioscoreaceae
10.	White yam	*Dioscorea alata*	Dioscoreaceae
Class Dicotyledoneae			
11.	New Zealand spinach	*Tetragonia expansa*	Aizoaceae
12.	Leaf beet	*Beta vulgaris var. Bengalensis*	Chenopodiaceae
13.	Spinach	*Spinacia oleracea*	Chenopodiaceae
14.	Beetroot	*Beta vulgaris*	Chenopodiaceae
15.	Lettuce	*Lactuca sativa*	Compositae
16 .	Artichoke	*Cynara scolymus*	Compositae
17.	Sweet potato	*Ipomoea batatas*	Convolvulaceae
18.	Cabbage	*Brassica oleracea var. capitata*	Cruciferae
19.	Cauliflower	*Brassica oleracea var. botrytis*	Cruciferae
20.	Brussels sprouts	*Brassica oleracea var. gemmifera*	Cruciferae
21.	Sprouting broccoli	*Brassica oleracea var. italica*	Cruciferae
22.	Kale	*Brassica oleracea var. acephala*	Cruciferae
23.	Kohlrabi or knol-khol	*Brassica oleracea var. gongylodes*	Cruciferae
24.	Rutabaga	*Brassica napus var. napobrassica*	Cruciferae
25.	Turnip	*Brassica campestris var. rapa*	Cruciferae
26.	Chinese cabbage	*Brassica chinensis*	Cruciferae
27.	Radish	*Raphanus sativus*	Cruciferae
28.	Pumpkin	*Cucurbita moschata*	Cucurbitaceae
29.	Summer squash	*Cucurbita pepo*	Cucurbitaceae
30.	Winter squash	*Cucurbita maxima*	Cucurbitaceae
31.	Watermelon	*Citrullus lanatus*	Cucurbitaceae
32.	Round melon	*Praecitrullus fistulosus*	Cucurbitaceae
33.	Cucumber	*Cucumis sativus*	Cucurbitaceae
34.	Muskmelon	*Cucumis melo*	Cucurbitaceae
35.	West Indian gherkin	*Cucumis anguria*	Cucurbitaceae
36.	Long melon	*Cucumis melo var. flexuosus*	Cucurbitaceae
37.	Ridge gourd	*Luffa acutangula*	Cucurbitaceae
38.	Sponge gourd	*Luffa cylindrica*	Cucurbitaceae
39.	Bottle gourd	*Lagenaria siceraria*	Cucurbitaceae
40.	Bitter gourd	*Momordica charantia*	Cucurbitaceae
41.	Pointed gourd	*Trichosanthes dioica*	Cucurbitaceae

42.	Snake gourd	*Trichosanthes anguina*	Cucurbitaceae
43.	Chayote or cho-cho	*Scchium edule*	Cucurbitaceae
44.	Ivy gourd	*Coccania grandis*	Cucurbitaceae
45.	Tapioca or cassava	*Manihot esculenta*	Euphorbiaceae
46.	Pea	*Pisum sativum*	Fabaceae
47.	French bean	*Phaseolus vulgaris*	Fabaceae
48.	Lima bean	*Phaseolus lunatus*	Fabaceae
49.	Broad bean	*Vicia faba*	Fabaceae
50.	Cowpea	*Vigna unguiculata*	Fabaceae
51.	Asparagus bean	*Vigna unguiculata var. sesquipedalis*	Fabaceae
52.	Cluster bean	*Cyamopsis tetragonoloba*	Fabaceae
53.	Dolichos bean	*Dolichos lablab*	Fabaceae
54.	Winged bean	*Psophocarpus tetragonolobus*	Fabaceae
55.	Soybean	*Glycine max*	Fabaceae
56.	Fenugreek	*Trigonella foenum-graecum*	Fabaceae
57.	Okra or lady's finger	*Abelmoschus esculentus*	Malvaceae
58.	Rhubarb	*Rheum rhaponticum*	Polygonaceae
59.	Potato or Irish potato	*Solanum tuberosum*	Solanaceae
60.	Eggplant or brinjal	*Solanum melongena*	Solanaceae
61.	Tomato	*Solanum lycopersicum*	Solanaceae
62.	Chilli or hot pepper	*Capsicum annuum*	Solanaceae
63.	Bell pepper or sweet pepper	*C. annuum var. grossum*	Solanaceae
64.	Carrot	*Daucus carota*	Apiaceae
65.	Parsley	*Petroselinum crispum*	Apiaceae
66.	Celery	*Apium graveolens*	Apiaceae
67.	Parsnip	*Pastinaca sativa*	Apiaceae

C. SPICES CROPS

S.No.	English Name	Botanical Name	Family
1.	Ginger	*Zingiber officinale Rosc*	Zingiberaceae
2.	Turmeric	*Curcuma longa*	Zingiberaceae
3.	Pepper	*Piper nigrum*	Piperaceae
4.	Cardamom	*Elettaria cardamom*	Zingiberaceae
5.	Coriander	*Coriandrum sativum*	Apiaceae
6.	Cumin	*Cuminum cyminum*	Apiaceae
7.	Fenugreek	*Trigonella foenum-graecum*	Fabaceae

D. MEDICINAL CROPS

S.No.	English Name	Botanical Name	Family
1.	Lemongrass	*Cymbopogon flexuosus*	Poaceae
2.	Citronella	*Cymbopogon winterianus*	Poaceae
3.	Palmarosa grass	*Cymbopogon martinii var.motia.*	Graminae
4.	Vetiver (Khus)	*Vetiveria zizanioides*	Poaceae
5.	Geranium	*Pelargonium graveolens L.Herit.*	Geraniaceae
6.	Davana	*Artemisia pallens Well.*	Asteraceae,

E. AROMATIC CROPS

S.No.	English Name	Botanical Name	Family
1.	Medicinal yam	*Dioscorea sp.,*	Dioscoreaceae
2.	Sarpagandha	*Rauvolfia serpentina*	Apocynaceae
3.	Opium	*Papaver somniferum L.,*	Papaveraceae
4.	Basilus	*Ocimum sanctum Linn.*	Lamiaceae
5.	Periwinkle	*Catharanthus roseus (L)*	Apocynaceae
6.	Aloe-vera	*A. barbadensis*	Liaceae
7.	Guggal	*Commiphora wightii (Arn.)*	Burseraceae
8.	Belladonna	*Atropa belladonna L.*	Solanaceae
9.	Medicinal solanum	*Solanum khasianum (Clarke),*	Solanaceae
10.	Senna	*Cassia angustifolia Vahl.*	
11.	Isubgol	*Plantago ovata Forsk.,*	Plantaginaceae
12.	Coleus	*Coleus forskohlii Syn.* *Coleus barbatus Brig.*	Lamiaceae

F. PLANTATION CROPS

S.No.	English Name	Botanical Name	Family
1.	Coconut	*Cocus nucifera*	Palmae
2.	Arecanut	*Areca catechu*	Palmae
3.	Betal vine	*Piper betle*	Piperaceae
4.	Cashew nut	*Anacardium occidentale*	Anacardiaceae
5.	Cocoa	*Theobroma cacao*	Sterculiaceae
6.	Coffee	*Coffea arabica* *C.canephora*	Rubiaceae
7.	Oil palm	*Elaeis guineensis*	Palmae

G. ANNUAL FLOWERS

S.No.	Common Name	Botanical name	Family	Flower Colour
1.	Floss flower	*Ageartaum houstonianum*	Compositae	Blue, pink, white
2.	Holly hock	*Althaea rosea*	Malvaceae	White, crimson, pink, yellow, red
3.	Snap dragon	*Antirrhinum majus*	Scrophulraiaceae	White, yellow, pink, rose, mauve
4.	English daisy	*Bellis perennis*	Compositae	White, blue, light pink
5.	Pot marigold	*Calendula officinalis*	Compositae	Yellow, orange
6.	Aster	*Callistephus chinensis*	Compositae	White, pink, blue
7.	Annual chrysanthemum	*Chrysanthemum coronarium*	Compositae	White, yellow
8.	Cosmos	*Cosmos bipinnatus*	Compositae	White, pink, yellow, mauve, magenta
9.	Dahlia	*Dahlia variabilis*	Compositae	Yellow, red, white, mauve, crimson
10.	Carnation	*Dianthus caryophyllus*	Caryophyllaceae	White, pink, crimson, mauve, yellow, violet, red
11.	Straw flower	*Helichrysum bracteatum*	Compositae	Yellow
12.	Petunia	*Petunia hybrida*	Solanaceae	Red, blue, white, pink, violet
13.	Star flower	*Phlox drummondii*	Polemoniaceae	White, red, blue, violet, pink
14.	Sage	*Salvia splendens*	Labitae	Scarlet, pink, orange, white, yellow
15.	African marigold	*Tagetes erecta*	Compositae	Yellow, orange
16.	French marigold	*Tagetes patula*	Compositae	Red, Orange, yellow
17.	Verbena	*Verbena hybrida*	Verbanaceae	Purple, pink, blue, red, white
18.	Pansy	*Viola tricolor*	Violaceae	Yellow, blue, red, purple, violet and bicolour

H. FLOWERS CROPS

S.No.	English Name	Botanical Name	Family
1.	Rose	*Rosa hybrida*	Rosaceae
2.	Jasmine	*Jasminum spp.*	Oleaceae
3.	Chrysanthemum	*Dendranthema grandiflora*	Asteracea
4.	Crossandra	*Crossandra spp.*	Acanthacea
5.	Marigold	*Tagetes spp.*	Asteracea
6.	Tuberose	*Polianthes tuberosa L.*	Amaryllidaceae

I. SHRUBS

S.No.	Common Name	Botanical name	Family	Colour of the flowers
1.	Acalypha	*Acalypha hispida*	Euphorbiaceae	Long drooping spikes of crimson red flowers and the foliage is copper coloured and variegated.
2.	Adenium	*Adenium obesum*	Apocynaceae	Succulent stem and flowers is funnel shaped. Slow growing plant ideal for pots.
3.	Allamanda	*Allamanda grandiflora*	Apocynaceae	Yellow funnel shaped flowers
4.	Artabotrys	*Artabotrys odoratissimus*	Annonaceae	Scented flowers, green and turn to yellow on ripening
5.	Aralia	*Aralia spp.*	Araliaceae	Quite hardy plants and can be grown in plains.
6.	Barleria	*Barleria cristata*	Acanthaceae	Flowers are pink, violet, yellow in colours
7.	Beloperone	*Beloperone guttata*	Acanthaceae	Terracotta red orderly arranged showy bracts
8.	Bauhinia	*Bauhinia tomentosa*	Caesalpinaceae	Sulphur yellow in colour
9.	Peacock flower	*Caesalpinia pulcherrima*	Leguminosae	Bushy shrub with few scattered prickles. Shrub gives orange scarlet flowers.
10.	Camellia	*Camellia japonica*	Theaceae	Double flowers with white, rose or pink colour
11.	Raat ki Rani	*Cestrum nocturnum*	Solanaceae	Scented flowers at night
12.	Dinka Raja	*Cestrum diurnum*	Solanaceae	Evergreen plants, bears white flowers during day time
13.	Clerodendron	*Clerodendron inerme*	Verbenaceae	White flowers
14.	Cassia	*Cassia alata*	Leguminoase	Yellow flowers during winter
15.	Pigeon berry	*Duranta plumieri*	Verbenaceae	Blue flowers
16.	Eranthemum	*Eranthemum bicolour*	Acanthaceae	Plants with colourful foliage and grows well in shady and semi shady areas
17.	Hamelia	*Hamelia patens*	Rubiaceae	Orange red flowers
18.	Ixora	*Ixora parviflora*	Rubiaceae	Bears large terminal trusses of flowers in white, yellow, pink, orange, scarlet and orange scarlet colour.
19.	Lantana	Lantana camera	Verbanaceae	Common weed but can be grown as shrubbery along paths and roads
20.	Lal Kaner	Nerium indicum	Apocynaceae	Bushy shrub with cane like stems. Bears sweet scented flowers and are white, pink deep red and cream in colour.

21.	Star cluster	Pentas cornea	Rubiaceae	Red, pink or violet colour
22.	Poinsettia	Poinsettia pulcherrima	Euphorbiaceae	Red or orange colour
23.	Tecoma	Tecoma stans	Bignoniaceae	Yellow colour flowers
24.	Pili Kaner	Thevetia Peruviana	Apocyanaceae	Yellow colour flowers

J. ORNAMENTAL TREES

S.No.	Common Name	Botanical name	Family	Features
1.	Garlic vine	*Adenocalymma alliaceum*	Bignoniaceae	Heavy evergreen climber with shiny leathery foliage and large trumpet shaped pink flowers blooming during March to June
2.	Morning glory	*Ipomea learii*	Convolvulaceae	
3.	Coral vine	*Antigonon leptopus*	Polygonaceae	Tuberous rooted quick growing climber, bearing rose coloured flowers in great abundance. Flowers throughout the year.
4.	Kanthali Champa	*Artabotrys odoratissimus*	Annonaceae	Large woody climber or half scandent shrub. Flowers are solitary, greenish yellow in colour when ripe and gives a strong scent resembling that of
5.	Nepal Trumpet climber	*Beaumontia grandiflora*	Apocynaceae	Quick growing ornamental climber with woody stem and large shiny foliage. Gives out white trumpet shaped lily like flowers with faint scent.
6.	Clerodendron	*Clerodendron splendens*	Verbanaceae	Dwarf growing climber with dark green leaves and bears dazzling crimson coloured flowers.
7.	Butterfly pea creeper	*Clitoria ternatea*	Leguminosae	Summer growing perennial climber with a woody base, can climb tall grasses and crops. Leaves are elliptic to narrowly lanceolate and Flowers solitary, attractive, deep blue, occasionally pure white,
8.	Railway creeper	*Ipomea cairica*	Convolvulaceae	Quick growing climber with palmate light green colour. Produces purple coloured flowers with large corolla tube.
9.	Blue potato creeper	*Solanum seaforthianum*	Solanaceae	

10.	Purple wreath	*Petrea volubilis*	Verbanaceae	Woody shrubby climber with wreath like spikes of star shaped light mauve flowers during summer
11.	Bridal bouquet	*Porana paniculata*	Convolvulaceae	Strong shrubby climber having large dark green heart shaped leaves and white fragrant small flowers in abundance
12.	Rangoon creeper	*Quisqualis indica*	Combretaceae	Widely grown and hardy climber. Curved spines enable it to climb and flowers are white at the time of opening but gradually turn into pink and reddish and hence white and pinkish flowers are seen together
13.	Heavenly blue , clock vine	*Thunbergia grandiflora*	Acanthaceae	Extensively growing heavy climber with dense foliage and lovely bell shaped blue coloured flowers with yellow shades inside it.
14.	Vernonia	*Vernonia amygdalina*	Asteraceae	Quick growing evergreen limber and useful for screening. The branches are pendulous hang close to their support thus forming a neat screen. The flowers are white in colour but insignificant
15.	Bougainvillea	*Bougainvillea spp.*	Nyctaginaceae	Versatile plant and can be used like bush, standard bush, climber, hedge, pot plant, bonsai, ground cover for sloppy land.

L. Palms

S.No.	English Name	Botanical Name
1.	Bottle palm	*Hyophorbe legenicaulis*
2.	Royal palm	*Roystonea regia*
3.	Fish tail palm	*Caryota urens*
4.	Chinese palm	*Livingstonia rotundifolia*

Plant Identification Key

A. FRUIT CROPS

Common name - ....

Botanical name - ....

Family - ....

Plant nature - ....

Type of fruit - ....

Type of inflorescence - ....

Other characteristics - ....

B. VEGETABLE CROPS

Common name - ....

Botanical name - ....

Family - ....

Plant nature - ....

Type of fruit - ....

Type of inflorescence - ....

Other characteristics - ....

C. SPICES CROPS

Common name - ....

Botanical name - ....

Family - ....

Plant nature - ....

Type of fruit - ....

Type of inflorescence - ....

Other characteristics - ....

Use - ....

D. MEDICINAL AND AROMATIC CROPS

Common name - ....

Botanical name - ....

Family - ....

Plant nature - ....

Type of fruit - ....

Type of inflorescence - ....

Other characteristics - ....

Use - ....

E. PLANTATION CROPS

Common name - ....

Botanical name - ....

Family - ....

Plant nature - ....

Type of fruit - ....

Type of inflorescence - ....

Other characteristics - ....

Use - ....

F. ANNUAL FLOWERS / FLOWERS CROPS

Common name - ....

Botanical name - ....

Family - ....

Plant nature - ....

Leaf size & leaf colour - ....

Flower colour - ....

Other characteristics - ....

Use - ....

G. SHRUBS

Common name - ....

Botanical name - ....

Family - ....

Plant nature - ....

Leaf size & leaf colour - ....

Flower colour - ....

Other characteristics - ....

Use - ....

H. ORNAMENTAL TREES

Common name - ....

Botanical name - ....

Family - ....

Plant nature - ....

Flowering time - ....

Flower colour - ....

Other characteristics - ....

Use - ....

I. CLIMBERS

Common name - ....

Botanical name - ....

Family - ....

Plant nature - ....

Leaf size & leaf colour - ....

Flower colour - ....

Other characteristics - ....

Use - ....

J. PALMS

Common name - ....

Botanical name - ....

Plant nature - ....

Other characteristics - ....

❐❐❐

Exercise No. 3 **Date : ____________**

PREPARATION OF SEED BED/NURSERY BED

Nursery bed

A nursery bed is a specially prepared portion of land put aside for raising seedlings. It acts as a temporary home for young plants until they are eventually planted in a permanent garden.

Type of nursery: Three types of vegetable nurseries:

1. Flat seedbed
2. Raised seedbed and
3. Sunken seedbed

1. Flat seedbed

Flat beds are used where water availability is adequate and there are no drainage problems. The field is divided into small size plots. Ridges are prepared around each bed which facilitate the cultural practices. In between two row of the beds, central irrigation channel is prepared through which each bed is connected. This is very simple and easy to practice.

2. Raised seedbed

Raised beds specially useful for raising seedlings during rainy season when stagnation of water becomes problematic and causes damping-off disease. Raised bed of 15 cm height from ground level is prepared. All the stump, stones, pebbles etc. are removed from the bed and FYM at the rate of 10 kg per sq. meter should be well mixed in the soil. In between two rows, a space of 50-60 cm is left so as to carry out cultural practices easily. The seed are sown in lines over the bed.

3. Sunken seedbed

This type of bed is useful and prepared mostly during winter season. This type of nursery is prepared 10-15 cm downward from the soil surface. The air blows across the surface of soil and the seedling in sunken nursery could be is saved from the hit of the cool breeze of the air .Furthermore, covering of sunken bed with polythene sheets become easy which is required for protecting the seedling from cool air.

1. Flat Seed bed

2. Raised Seed bed

3. Sunken Seed bed

Inputs: Seed, khurpa, spade, rope, tape, watering can, hoe, rake and fork

Preparation of nursery bed

Site selection:

The nursery should be located near the planting site so as the transplants are taken to the main garden without any damages where as distances are long. Site should be located near a water source because irrigation is necessary. The site should be protected from strong wind, should not be located on a slope unless it is terraced since this may lead to erosion of not only the necessary top soil but the plants as well. Site should located where ample sunshine is available.

Bed preparation

At the time of field preparation while digging all weeds and stables are removed. The size of bed should be one meter wide and of convienent length. Fertilizer each three meters square with one wheelbarrow of compost and sand to make a thick layer then water the bed.

Seed rate

Type of vegetable	Seed rate (hac)
Tomato	125-175 g
Chilli	1.5 kg
Brinjal	200 g
Onion	10-12 kg
Cauliflower	500-600 g

Seed sowing:

Seed should be sown in rows. The rows should be prepared in east-west direction. Seed should be sown in rows and cover the seed with a thin layer of soil 2 cm deep at a spacing of 7 cm aparts. Mulch the bed with dry grass, and then using a watering can. Light irrigation should be done as an when required.

Germination:

The seeds start germinating after 5 – 7 days after sowing. Remove the grass mulch so that seedling starts growing.

Transplanting:

When seedling are ready for transplanting remove the shade a day before transplanting as this gives the seedling chance to get used to the strong sun-shine. Seedling transplants late in the evening (from 4 – 6 pm).

Care of nursery:

1. Regular watering of nursery should be done.
2. From day 1 of sowing, mulch your nursery and water it immediately after sowing.
3. After 5 to 7 days, check the germination of seed.
4. Continue watering on same frequency until the seedlings have at least three to four true leaves.
5. Depending on the weather, spray the nursery plants with a mild or curative fungicide like Carbendagem/Indofil M-45.

Fill the table

Type of vegetable	Duration to Seed Germination	Max Duration to Transplanting of Seedling
Capsicums	....	....
Tomato	....	....
Brinjal	....	....
Onion	....	....
Cauliflower	....	....

]

❐❐❐

Exercise No. 4 **Date : ____________**

PRACTICE OF SEXUAL AND ASEXUAL METHODS OF PROPAGATION

ASexual propagation

Sexual reproduction refers to multiplication of plants by seed. In sexual process, male and female gametes are fertilized to produce seeds.

Seed

Botanically seed is fertilized ovule. The true seed is a plant part having the following three Component: a living embryo, food resource or endosperm and protective coat.

eg. Papaya

Materials

Jamun, Mangosteen, Ber and Guava Seeds (old, new, water-soaked, TTC-treated), razor blades, petri dish, paper towels, sand a paper or file.

Procedures

1. **Seed Anatomy**
   Cut through a soaked bean seed and observe the internal structure. Sketch and label the parts of the seed.
2. **Seed Viability**
   Bean seeds have been soaked overnight in TTC (triphenyltetrazolium chloride). This changes living tissue to a red color. Uncolored spots will indicate poor viability. Cut open several seeds and sketch your observations.
3. **Seed Germination Tests**
   Seed has been divided into "old" and "new" lots. Count out 10 seeds from each lot and plant according to instructor's directions. Record the number of seeds that germinated for each group and calculate corresponding germination percentages.
4. **Seed Scarification**
   Physical or chemical abrasion of seed coat.This exercise will evaluate scarification techniques and their effect on germination percentages. Select 10 seeds for each of the four treatments and plant in the four different containers provided.

5. Seed Stratification

The term "moist chilling" synonym to "stratification"

(a) Treatment 1- Control (no scarification / no stratification)

(b) Treatment 2- Seeds soaked in hot water

(c) Treatment 3- Seeds soaked in acid (sulphuric acid)

(d) Treatment 4- Mechanical scarification (use sandpaper, file, or clippers)

(e) Treatment 5- Moist chilling (Stratification)

IV. Results

Obtain seed germination data for the steps 3 and 4 above for your group and the entire class. Use this information for your lab report.

S. No.	Crop Name	Treatment	Number of Seed Viability	Number of Seed Germination
1.	Guava			
2.	Walnut			
3.	Mangosteen			
4.	Ber			
5.	Jamun			
6.	Peach			

B. Asexual propagation

Asexual propagation involves no change in genetic make-up of the new plant. All the characteristics of the parent plant are reproduced in the daughter plant due to exact duplication of chromosome system that takes place during cell division. Thus, the plant true to type in growth, ripening, yield and fruit quality. The various method of asexual propagation are division, cutting, layering, budding and grafting.

1. Division :

It is a method of asexual propagation by which vegetative part that separate naturally from the parent plant or may be readily separated from mother plant, is used.

(i) Rhizomes:

Botanically, a rhizome is an elongated, horizontal fleshy underground stem having nodes and internodes bearing buds in the axils of reduced Scale leaves. The rhizome are cut into pieces, each with one or two viable bud and planted in favourable atmospheric conditions. Adventitious roots and shoots develop from the nodes. e.g. Banana

(ii) Suckers:

A shoot arising on an old stem or underground part of the stem or on horizontal root systems is known as suckers. e.g. Dateplam, Banana and Pineapple

(iii) Offsets:

An offset is a characteristic type of lateral shoot or branch that develops from the base of the main stem in certain plants. e.g. Datepalm

1. Rhizome

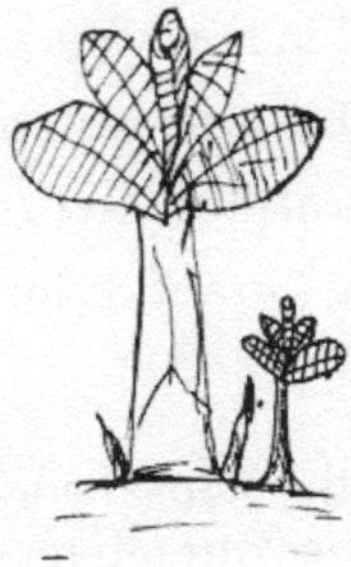

2. Banana Suckers

3. Offsets of Date Palm

(iv) Crown:

It is a very short condensed part of plant at the surface of the ground which gives rise to new shoots. Crown grown on top of the pineapple.

(v) Runners:

A runner is a specialized stem that develops from the axil of a leaf at the crown of a plant, grows horizontally along the ground and forms a new plant at one of the nodes. e.g. strawberry.

4. Pineapple Crown

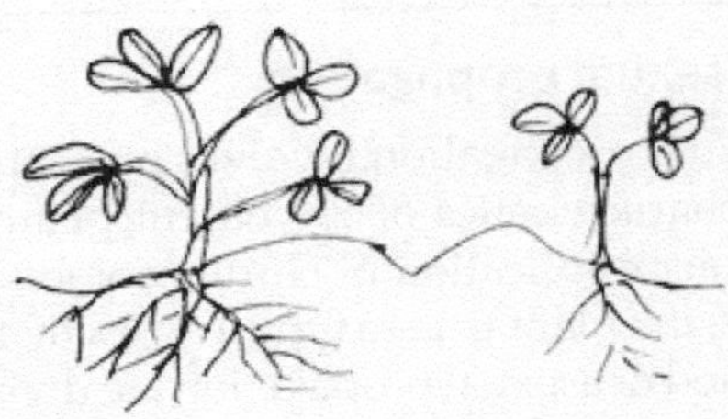

5. Strawberry Runners

2. Cutting:

Cutting is a detached vegetative part of a plant, which on separation and planting is able to regenerate the missing parts and develop itself into a new plant. It is an inexpensive and quick method of propagation. A large number of uniform plants can be produced using few parent plants. It does not involve specialised skills. The method is named after the part of plant used for cutting. e.g. stem, root and leaf.

Material required:

Branch of bougainvillea, lemon, chrysanthemum and bryophyllum leaf secateurs or sharp cutter, rooting hormone IBA, nursery bed

Stem cutting

Based on the age and maturity of shoots detached for vegetative propagation, stem cuttings is of four types.

(i) Hardwood cutting (ii) Semi-hardwood cutting (iii) Softwood cutting (iv) Herbaceous cutting

(i) Hardwood cutting

Select branches of bougainvillea from one-year old healthy plants, having pencil thickness. Cut the branches into 10–15 cm long cuttings. Each cutting must have at least 4–5 dormant vegetative buds. Leaves and thorns, if present, are completely removed. This checks transpiration loss. A straight cut is given at the base of the cuttings just below the node and a slanting upper cut is given away from the top bud. The cut portion will help identify the planting position. The secretion of hormones at the bud near the cut portion induces rooting. Slanting cut at upper end reduces transpiration loss, which can be inhibited by the application of wax. The cuttings are planted slant-wise in a nursery bed or small poly bags for growing plants. Callus tissues form the cambium layer and rooting takes place in this region. The best season for planting the cuttings is monsoon for evergreen plants and December-March for deciduous plants. Cuttings can be planted in greenhouse or poly-house for better results. e.g. rose, grapes, fig, pomegranate, bougainvillea, lagerstroemia, jasminum, hibiscus, etc.

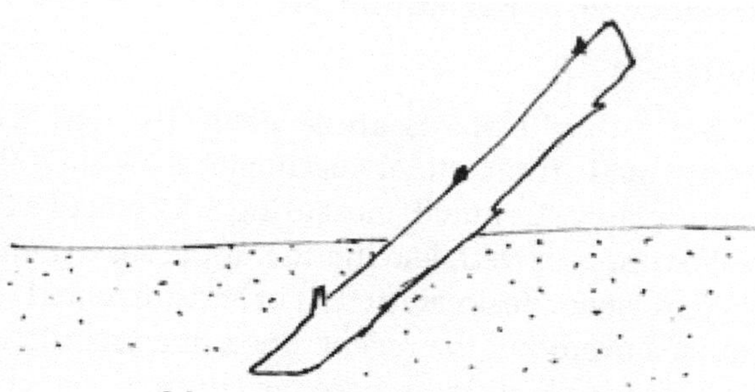

1. Hard wood cutting

(ii) Semi-hardwood cutting

A semi-hardwood cutting is taken from 4 to 9-month old shoots of current season woody plants and semi-hardwood cuttings are prepared from branches having pencil thickness. The length of these cuttings varies from 7.5 to 15 cm. The cuttings must have at least 4–5 dormant vegetative buds. Some leaves are retained as they help in preparing food by photosynthesis. Large leaves are reduced in size by cutting. A straight basal cut is given just near the vegetative bud and a slant top cut must be given away from the bud. Dipping the base of the cuttings before planting in IBA @ 5000 ppm induces early rooting. The cuttings are planted in slanting position so that their maximum base is in contact with the rooting medium. The planting season for semi-hardwood cuttings is monsoon. Commercially, such cuttings are rooted under mist spray or fog. e.g. croton, acalyphas, aralias, diffenbachia, russelia, cestrum, nerium, etc

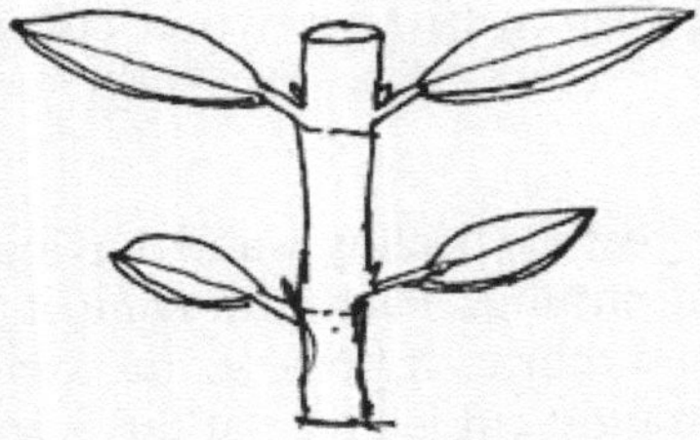

2. Semi hard wood cutting

3. Soft wood cutting

(iii) Softwood cutting

Shoots of 2 to 3-month old plants are selected for softwood cuttings. Softwood cuttings are prepared from tender but mature branches. The length of these cuttings varies from 10–12 cm. Tender shoots do not have sufficient food material. Hence, all leaves present on the shoots are retained for photosynthesis. The cutting material are gathered early in the morning and must be kept moist by keeping them in a wet cloth. Sandy loam medium is the best for planting softwood cuttings. e.g. alternanthera, coleus, duranta, clerodendrum, etc.

(iv) Herbaceous cutting:

Shoots of 1 to 2-month old plants are selected for herbaceous cuttings. Herbaceous cuttings are made from tender succulents, especially the leafy part of the stems of herbaceous plants. Terminal, measuring 8–12 cm, of a healthy shoot is cut and the basal leaves are removed, leaving the upper leaves undisturbed. The cuttings once detached must not desiccate at the cut and are rooted well under mist. The application of auxins promotes the regeneration of adventitious roots. Sandy loam medium is the best for planting herbaceous cuttings. e.g. chrysanthemum, iresine, pilea, dahlia, petunia, carnation, marigold, etc.

4. Herbaceous Cutting

Root cutting

The root cuttings should be taken from root pieces of young stock plants in late winter or early spring when the roots are well supplied with stored foods but before new growth starts. It is important with root cuttings to maintain the correct polarity when planting as the new shoot develops from the proximal end i.e. from the part close to the crown. The proximal end of the root piece should always be up. e.g. apple, pear, cherry, guava, wood apple etc.

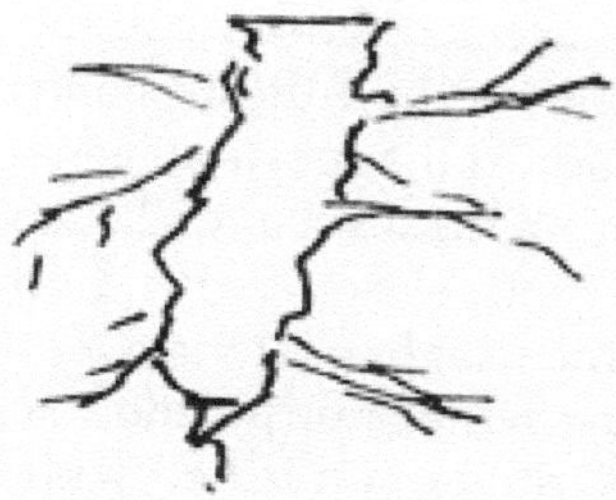

Root Cutting

Leaf cutting

Plants with thick fleshy leaves having buds are propagated by leaf cutting. Vegetative buds are present in the notches of leaf margin (bryophyllum) or on the vein (begonia rex). Leaf blade or pieces of it with bud are put on the rooting medium under favourable conditions. In case of black raspberry, the leaf blade, along with petiole and a short piece of the stem with attached axillary buds, are kept in the medium for rooting. e.g. snake plant (senseveria), blackberry, rhododendron and bryophyllum

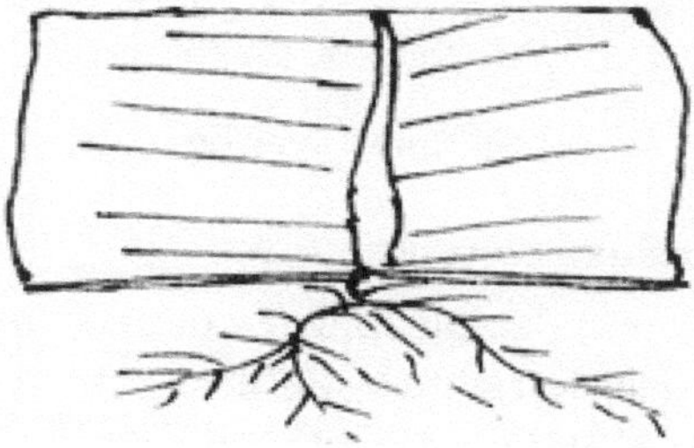

Leaf Cutting

3. Layering:

It is an attached method of propagation. In this method, roots are allowed to develop on the covered portion of the stem while still being attached to the mother plant. After the emergence and development of the roots, this portion is separated from the mother plant and allowed to grow as a new plant on its own root stem. Such root stem is known as 'layer'.

Types of layering

(i) Simple layering (ii) Compound or serpentine layering (iii) Trench layering (iv) Mound layering or stooling (v) Air layering

Material required:

Sharp knife, stone pieces or hooks or pegs, polythene bags, secateurs and matchstick

(i) Simple layering

Select one-year old healthy, flexible, long un-branched shoot near the ground level. Remove leaves of the selected shoot, retaining some at the top. The retained leaves prepare food through photosynthesis. Bend down the shoot so that some part of it touches the ground. At that portion, generally, 15 30 cm away from the terminal end, a sharp slanting inward cut of 2 3 cm is given. A small matchstick is inserted in the cut to keep the slit open. Bend down the branch and cover the cut part with soil. Keep some weight or stone over the buried part so that it is not pulled upward, and remains in the same position. A stake is fixed near the layered branch and the branch operated upon is tied with it. Water the layered portion regularly. After 3 4 weeks, rooting starts at the operated portion and this can be indicated by sprouting buds on the shoot. After this, the layer is separated from the mother plant and planted in a new place. e.g. jasmine, ixora, clerodendron, pyrostegia, etc.

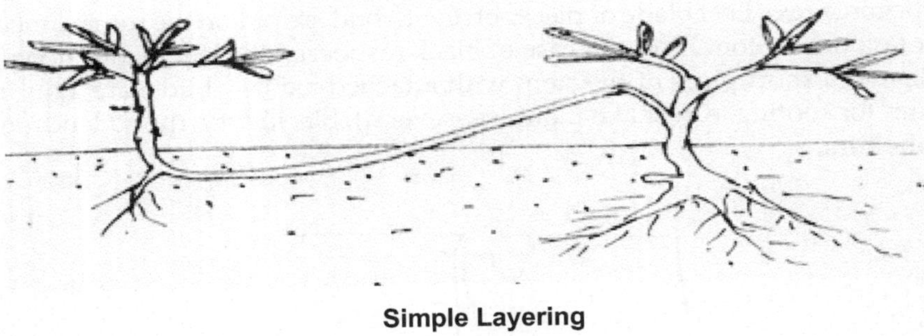

Simple Layering

(ii) Compound or serpentine layering

One-year old healthy and flexible long shoot near the ground is selected for compound layering. The selected stem is placed in soil in a way that the nodes at certain distance are covered under the soil and the intermediate internodes are exposed. Remove leaves from the selected branch but retain few leaves at the top. Give two circular cut around the bark about 2.5 4 cm wide. Remove the bark of the operated portion (girdling). Apply rooting hormone to the girdled portion and cover it with soil. The same branch is operated at 3 4 places at certain distance in the same way. The growing shoots, which emerge from the covered portion of the branch, are separated from the mother plant for planting in a nursery. e.g. bougainvillea, jasmine, clematis and muscadine grape etc.

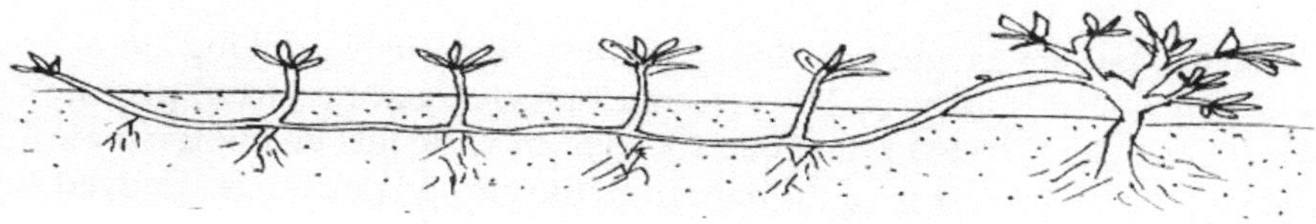

Compound or Serpentine Layering

(iii) Trench layering

One-year old healthy and flexible long shoot near the ground is selected. The selected stem is placed in a shallow trench in a way that the middle portion of it is buried. Remove leaves from the branch but retain few leaves at the top. Cover the whole branch with moist soil 5–10 cm deep. The terminal portion is left exposed to manufacture food and hormones for the developing plants. After some weeks, shoots arise from the nodes, which are covered by soil. The covering of the shoots with soil results in etiolation of the shoots and helps in rooting. Individual shoots with roots (layers) are separated from the mother plant and planted in a nursery. apple, cherry, pear, jasmine and rhododendron.

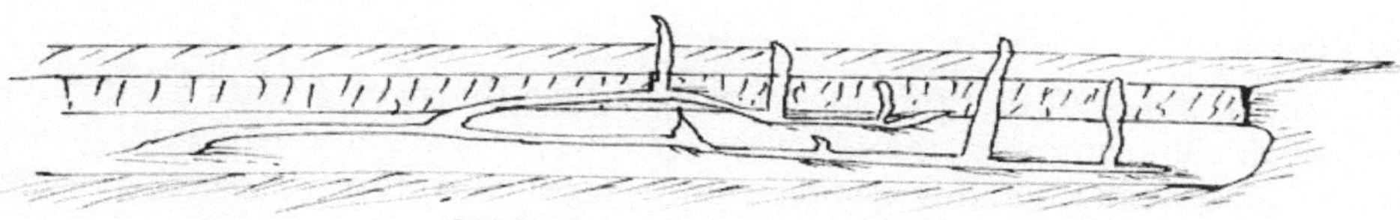

Trench Layering

(iv) Mound layering or stooling

Cut back the upper portion of the plant 2.5 cm above the ground level. After few days, new shoots will emerge. When the shoots grow to a height of 7 15 cm and become little sturdy, place loose soil around them so that they are half buried. When the shoots attain a height of 20 25 cm, again add soil around them so that they are half buried. Water the heaped soil regularly. It will take 3 4 months to get the layers. Cut the rooted layers close to the base from the mother plant and plant it in a nursery. e.g. are apple, guava, currant, gooseberry, pear, etc.

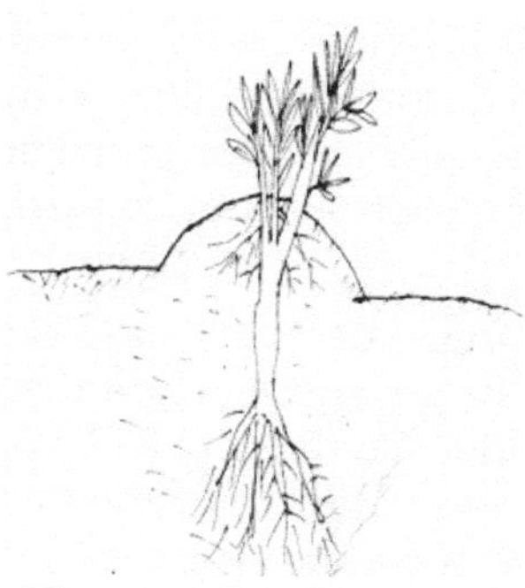

Mount or Stool Layering

Air Layering

(v) Air layering

It is also known as 'gootee'. Select healthy, vigorously growing aerial branch having pencil-size thickness. The selected branch must be of the past growing season. Girdle the selected branch up to 2 3.5 cm wide just below the node 15–30 cm prior to the shoot. A strip of the bark from the girdled portion is removed. Scrap the girdled portion, which helps in the removal of phloem tissues and prevents formation of bark at the girdled portion. Excessive moisture from sphagnum moss is squeezed out before placing it over the cut portion. A piece of polyethylene film is carefully wrapped around the branch so that the sphagnum moss is completely covered. Both the ends of the polyethylene film are made airtight by tying them with strings. The layer is removed from the parent plant when roots are observed through the transparent polyethylene film. It takes 2 3 months for rooting. Rainy season is the best for air layering. e.g. Ficus elastica, Callistemon, Croton, Monstera, Citrus fruits, Litchi, and pomegranate, etc.

4. Budding

Budding is the process of inserting a single mature Scion bud into the stem (rootstock) in the way that results into a union and continues to grow as a new plant.

Rootstock

The part of the graft that provides root system to the grafted plant is known as 'rootstock'. It is, normally, raised by seeds in the seedbed, and then, transplanted in the nursery bed for budding and grafting. Rootstocks are also raised in pots and polythene bags.

Scion

The upper portion of graft combination taken from the desired plant to be multiplied is known as 'scion'.

Types of budding

(i) T – budding (ii) Patch budding (iii) Ring budding (iv) Flute budding (v) Forkert budding (vi) Chip budding

T – Budding

Since a 'T'-shaped incision is made for bud insertion on the rootstock, it is called T– budding. T / Shield budding is also called 'shield budding' as the bud used for insertion is in the shape of a 'shield'. It is widely used for propagating fruit trees and ornamental plants. In this method, the rootstocks of compatible plants are raised in beds or poly bags.

One-year old rootstock of a healthy and vigorous growth is selected. A T-shaped cut is made at a height of 15 25 cm from the ground level on the rootstock with the help of a sharp knife. Also, a vertical cut, extending up to 2.5 3.75 cm is given on the bark in the internodes. At the top of this vertical cut, another horizontal cut is given in such a way that the two cuts resemble the shape of 'T'. Now, the bark of the cut portion is loosened. The bud from the desirable plant is selected. The shield-shaped bud along with the woody chip is removed from the scion stick. After remaining the chip of bud the bud is inserted in the T-shaped cut in the stock. After the insertion of the bud with the help of a budding knife, the bud and stock are firmly wrapped with a polyethylene strip, exposing the bud. The bud sprouts within a month. Examples are rose, apple, pear, peach, apricot, cherry, sweet orange, etc.

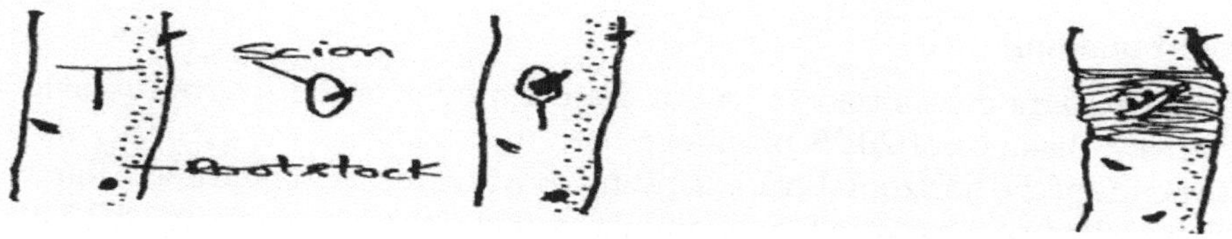

T-budding

Patch budding

A rectangular patch of bark, measuring 2.4 1.5 cm (length and width), is completely removed from the internodes of the stock plant. A similar patch of bark with a healthy bud is removed from the scion bud stick. This patch is placed on the cut portion of the stock and wrapped with a polyethylene strip, keeping the bud exposed. This type of budding is useful for the propagation of plants having a thick bark. Examples are aonla, mango, jamun, rubber, etc.

Patch budding

Modified Ring budding

In this method, a bark of approximately 3 6 cm wide in ring form is removed from the stock. The same dimension of bark with a healthy bud is removed from the scion bud stick and placed on the stalk. After placing the ring in position, tie it with a polythene strip, keeping the bud exposed, e.g. Ber and Cherry.

Ring budding

Flute budding

This is a slight modification of ring budding. Instead of removing the complete ring, a narrow portion of the bark about 1/8 of its circumference is left on the stock. A similar portion of the scion is removed along with the bud and is fitted on the cut portion of the stock. The bark of the stock and bud are tied with a polyethylene strip, exposing the growing point e.g. ber.

Forkert budding

In forkert method, a horizontal cut at the internodes of the selected rootstock is given at a distance of 20 25 cm above the ground level. Two vertical cuts from either ends of the horizontal cut, extending downwards, are taken and a flap of the bark is pulled out, exposing a rectangular woody portion of about 2.5 2.5 cm on the rootstock. A rectangular piece of bark of the same size along with a matured bud is removed from the bud stick with the help of a budding knife. This piece of bark with bud is then shifted over the exposed portion on the rootstock. The flap of the bark is loosened and tied to its original position, covering the scion bud fitted inside. After three weeks, the polyethylene strip is removed and the flap of the bark is pulled out for observation. If the bud shows sprouting signs, the flap is removed by giving a horizontal cut on the downside. Then, the polyethylene strip is wrapped, keeping the growing point exposed. The bud sprouts within 3 – 5 weeks of budding operation. Examples are cashew nut, jackfruit, etc.

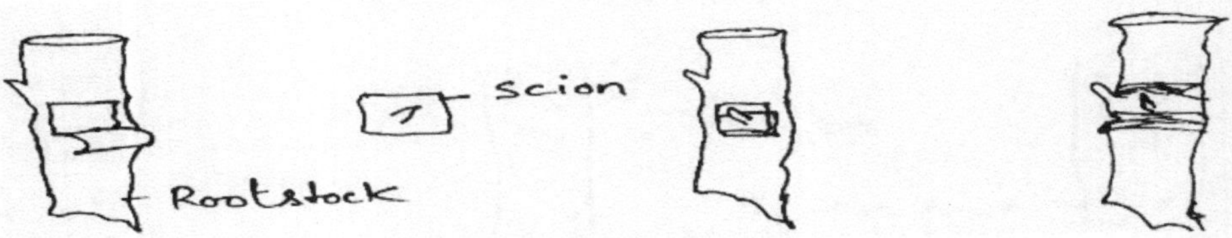

Forket budding

Chip budding

This method is followed when the bark is thin and cannot be removed easily. In this method, a piece of thin bark, along with some wood piece, is removed between two nodes of the rootstock, and the same size of chip, which is similar in shape and is collected from the scion, is placed on the rootstock. This is mostly practised in February March. Fruits like apple, grapes and pear can be propagated through this technique.

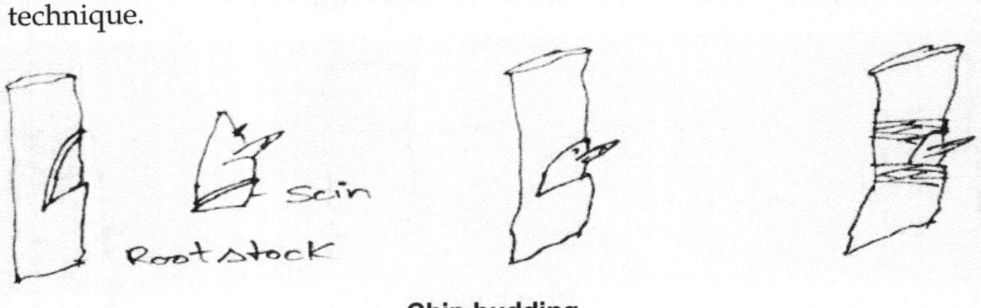

Chip budding

5. Grafting

The method of joining parts of two plants in a manner that they form a unit and function as one plant is known as 'grafting'.

Methods of grafting

Scion attached method

In this method, the scion shoot is not detached from the mother plant until the union takes place. After the successful union of the scion and rootstock, the scion is separated in gradual cut from the mother plant. For making the grafting handy, the rootstock is grown in a container or polythene bag. This method is followed in plants, in which successful graft unions are difficult to obtain. Approach grafting' is a type of scion attached method. It is classified into two types.

- Sliced approach grafting
- Tongue grafting

Approach grafting

Approach grafting is also known as 'inarching'. The main feature of approach grafting is that two independent self-sustaining plants are grafted together. After the successful union of the graft, the scion plant is detached below the graft union from the mother plant and the top of the rootstock plant is removed above the graft. This method is useful for plants, in which successful graft unions are difficult to obtain. This method is, usually, performed for plants growing in a container, as well as, big trees. In the latter case, the rootstock seedling is brought near the scion branch by erecting a platform.

Sliced approach grafting

Bring the selected rootstock and scion close together. Find out the most comfortable point of contact. At the point of contact, a thin slice of wood along with a 2.5 to 5-cm long bark from the rootstock and the scion is removed. The operated size must be uniform on both the stems of the rootstock and the scion. The cut surfaces are then brought together so that they cover each other completely by overlapping. Press them firmly together and tie them with a waxed string or polythene tape, so that water does not enter. After successful union, head back the rootstock above the union and cut the scion below the union, e.g., mango, guava, sapota, etc.

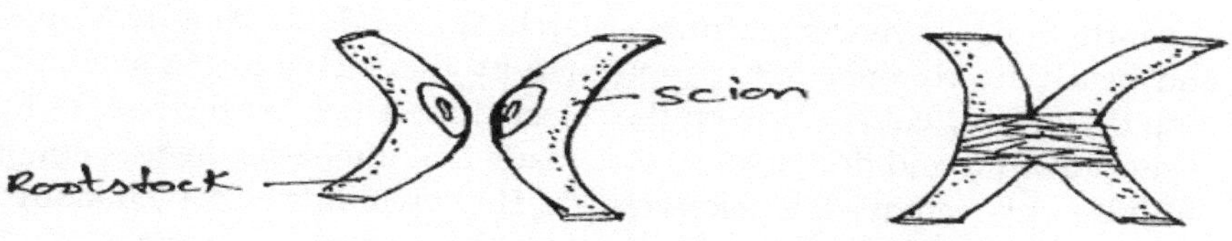

Sliced approach grafting

Tongue grafting

This method differs from the former as cuts are given on both the scion and rootstock. Bring the selected rootstock and scion close together. Find out the most comfortable point of contact. Remove a slice of wood along with a 2.5 to 5-cm long bark from the rootstock and scion. A second slanting partial cut downward on the stock and upward on the scion is made, producing a thin tongue-like structure of the same size on the stem of the stock and the scion. Insert the scion in the stock so that these tongue cuts interlock. All operated portions must be in contact with each other. Tie the operated portions.

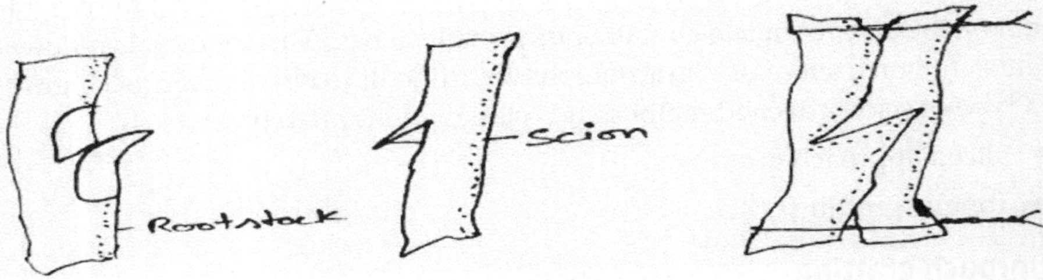

Tongue grafting

Scion detached method

This method is a more popular method of grafting and comparatively easier to perform. Besides, the rate of success of plant propagation is more in this case. The scion is first detached from the mother plant, and then, inserted or tied on the rootstock. For preparation scion, the scion stock must be defoliated one week before the grafting operation. The types of scion detached method are:

- Veneer grafting
- Side grafting
- Wedge or cleft grafting
- Stone or epicotyl grafting
- Whip or splice grafting
- Bark grafting

Veneer grafting

It is a simple and economical method of grafting. The most ideal method for establishing In-situ orchards and top working of old unproductive orchards. The best time in north India for veneer grafting is March April and July August. Mango, cashew and peach are commercially propagated by this technique. A shallow 3 to 5-cm long downward cut is made on the selected rootstock. At the base of the first cut, a short inward and downward cut is made that intersects the first cut. In between both the cuts, remove the piece of wood along with the bark by making a small notch in the rootstock. The scion is operated with a matching long cut on one side and a short cut on the opposite side is given at the base. Insert the scion and fix it in the rootstock. Care must be taken to ensure that the cambium layer matches at least one side of the cut surface. Wrap and tie the scion and rootstock firmly. Cut back the rootstock above the union after successful union.

Veneer grafting

Side grafting

In this method, the operated scion is inserted into the side of the established rootstock, which has more girth than the scion. A rootstock of 2.5 cm diameter is selected. The scion needs to have 3–5 buds and must be about 7.5 cm long. The scion must be comparatively thinner than the rootstock. Use a sharp grafting knife for cutting the scion. On the stem of the rootstock, a slanting downward and inward cut of about 2.5–5 cm deep is made. A wedge of the same size (2.5 cm) of the scion is prepared by two slanting cuts oppositely towards the base. The scion is then inserted into the operated rootstock. Pour wax and make the operated portion waterproof. Wrap and tie the grafted portion to keep it intact. After the graft union is complete, cut the stock above the union.

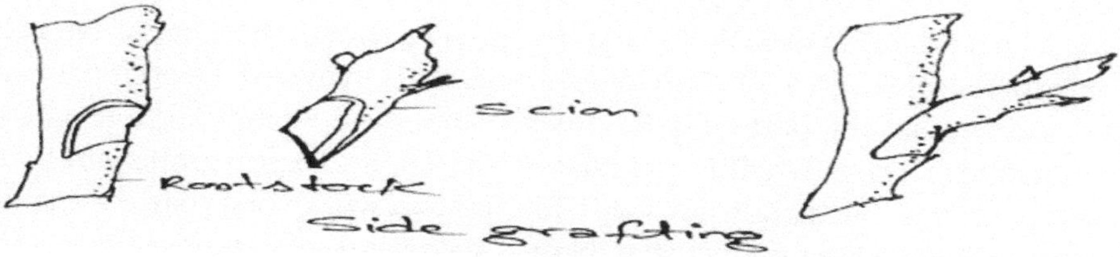

Side grafting

Cleft grafting

It is comparatively a simple and an easy method of grafting, which is widely used in fruit trees, e.g., mango, jackfruit, aonla, etc. The scion must be a terminal shoot with 3–5 buds. It must be of the current season and in active growth. The scion shoot is defoliated about one week ahead of being separated from the mother plant. This will help accumulate food in the shoots. As a result, the buds on the shoots become swollen. As compared to the rootstock, the thickness (diameter) of the scion may be the same or less. The rootstocks of required plant species are raised in poly bags. The seedling of the suitable rootstock, which is 4 to 5–month old is selected. Head back the rootstock. A sharp vertical downward cut of 3 5 cm is made in the centre of the stem. Two slanting cuts of the same length (3–5 cm) as in the rootstock are given on the sides towards the base on the scion shoot. This will give a wedge-shaped appearance to the scion stick. The wedge-shaped scion is inserted in the split of the rootstock. Insert the scion in a way that it matches the cambium layer at least on one side with the stock. Tie the grafted portion firmly in position with a polythene tape. After successful union, the terminal buds of the scion begin to sprout. Loosen or remove the polythene tape to allow the shoot to grow normally.

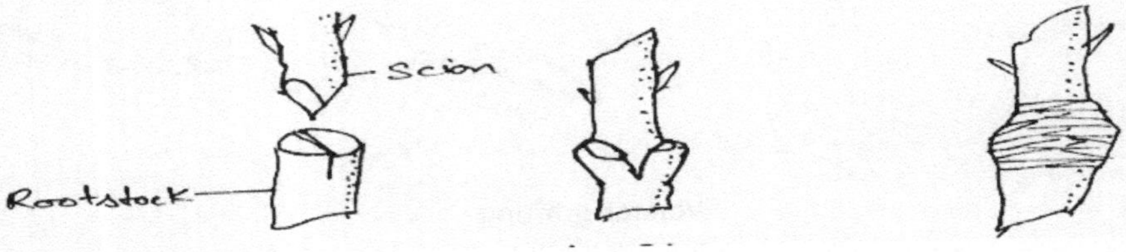

Cleft grafting

Stone or epicotyl grafting

This method is commonly adopted for the rapid multiplication of mango plants. In this method, stones (seeds) are sown in polythene bags or moist sand bed and covered with 5 to 7-cm layer of leaf mould for germination. When the seedlings are about 15 days old, they are taken out and grafted indoor. Young mango seedlings with copper coloured leaves (15 days old) are used as rootstocks. Young (current season growth) 3 to 4-month old shoot with pencil thickness is selected as the scion on the mother plant. The scion must have 4 5 terminal (apical) buds. Defoliate the selected shoot 10 days prior to grafting. Defoliation makes the buds swollen and induces early sprouting. A wedge in the scion is made at the base by giving two slanting cuts of 5 cm. Head back the stock by giving a straight horizontal cut. From the centre of the stock, give a 3 - 5cm long vertical cut downward. Insert the wedge-shaped scion in the split portion of the stock so that the operated portion is in full contact. Tie the graft firmly with a polythene strip. The successful graft sprouts and new shoots emerge. Such grafts are ready for planting within one year, e.g. mango. Such graft should be kept in mist chamber for successful sprouting.

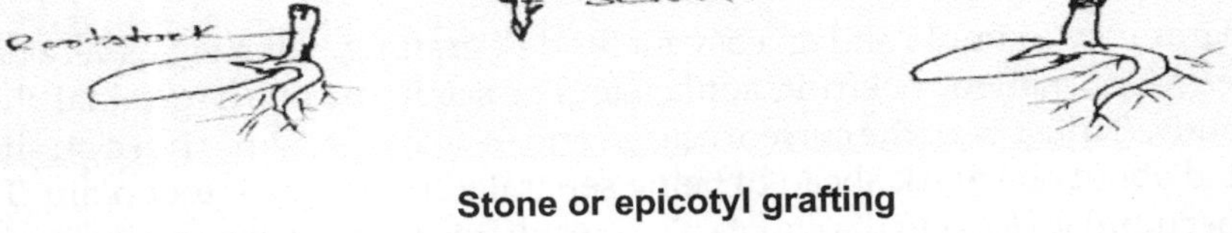

Stone or epicotyl grafting

Whip or splice grafting

It is the oldest method of grafting. This method is used in fruit trees like apple, pear, walnut, etc. Select one-year old rootstock. The rootstock and scion must be of uniform thickness. The scion must be 10 to 15 cm long having 4–5 swollen buds. The rootstock must be in active growth phase and sap-flowing condition. It is mostly performed in early spring season. Head back the rootstock terminally.

Give a slanting cut of 2.5 5 cm downwards from the top. Operate similarly but reversely on the scion. On the scion, a slanting cut of the same size is given from the base upward. The cuts on both the stock and scion need to be smooth. Put the operated portions on each other so that they form a single stem. Wrap the union with a polythene tape or a special nursery tape. The tape must be removed after the graft has healed, else the growth is restricted around the union, and such plants break due to the force of wind.

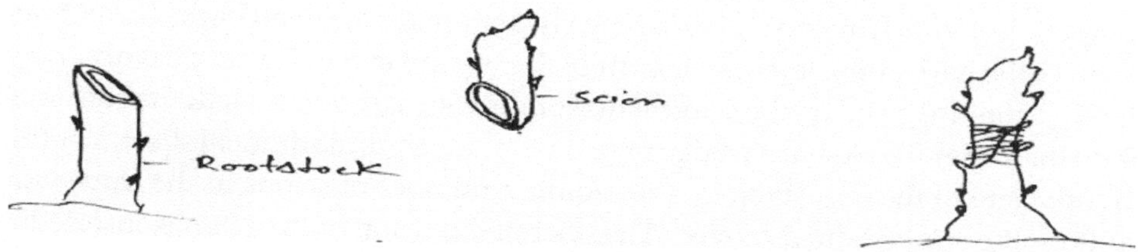

Whip or splice grafting

Bark grafting

A plant graft made by slitting the bark of the stock and inserting the scion beneath it is called 'bark grafting'. It is commonly used in top working. The bark of the rootstock must be in sap-flowing condition. The scion must be in dormant condition. The scion must be 10–13 cm long and have 3 5 dormant buds. A vertical cut of 3 5 cm is made in the bark of the stub of the rootstock. To prepare wedge shape of the scion, a 3 to 5-cm long cut is made at the end of the scion, followed by another cut on the opposite side of the first cut. Slightly lift the bark of the rootstock of the cut portion. Insert the wedge-shaped scion into the rootstock and cover it with the bark of the rootstock. After grafting, the exposed cut surfaces of the stub and scion are covered with wax. Several scions may be used for grafting on a single stock, according to the width of the stub.

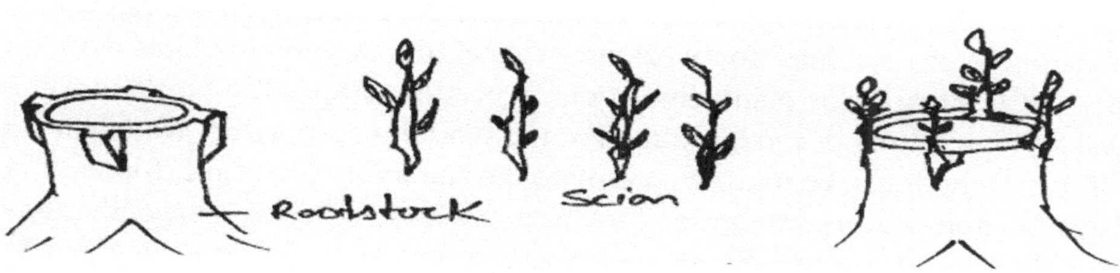

Bark grafting

Grafting for special purposes

Grafting for special purposes is done for quality improvement and repair. Grafting is done on established plants to improve their quality, e.g., side grafting and top working. It is also done to rejuvenate old or injured trees, for example bridge grafting.

Bridge grafting

This method is used for repairing wounds in trees made by implements, frost, rodents or diseases. In this grafting, the bark of a tree is damaged, resulting into girdling. A completely girdled tree will die. Bridge grafting repairs girdling. The rootstock must be in sap-flowing condition. The scion comprises one-year old dormant shoots of 6 12 mm in diameter. The number of scion sticks depends upon the size of the wound to be repaired. The selected scion may be of the same or a compatible plant. Trim the wounded area by removing the dead bark. The cuts are made in the bark at the top and bottom of the wound at 5 to 7.5 cm distance. Long slanting cuts are given on the scion at the top, as well as, bottom. Both the cuts must be on the same side. A sharp wedge of scion is made by an additional short, slanting cut opposite to the first. Prepare the required number of scions in the same way. Buds on the scion(s) are removed. The operated portion of the scion is inserted in each slot of the bark on the rootstock in a way that the wedge remains under the flap of the bark at each end. The scions must be put in upright position. The graft unions at top to bottom are waxed.

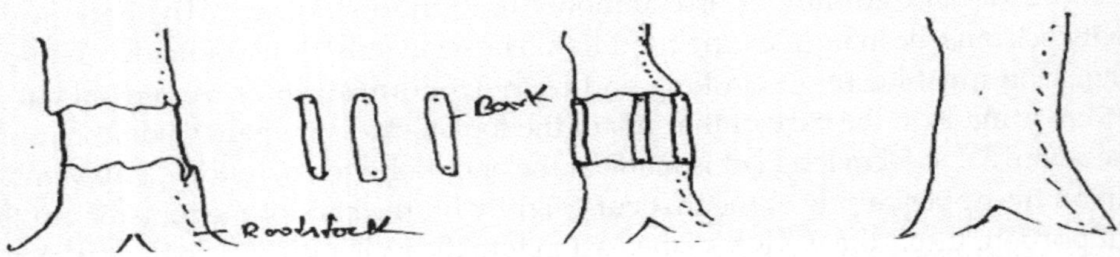

Bridge grafting

Top working

Top working is a method of grafting by which inferior or older plants are rejuvenated into superior or new ones. Top working is, generally, adopted in plants with long leaves. It is suitable for plants like apple, avocado, mango. Top working can be done by top grafting or top budding. For top working, cleft, whip, wedge or side grafting methods can be used, according to the suitability of a plant. Top working is, usually, done during spring. Prepare the rootstock and scion by any of the above suitable method. Usually, 3–5 scaffold branches must be used for rejuvenation in top working. In frame working, the secondary scaffold branches are used for grafting. The branches to be worked which must be well-distributed around the trunk. The branches must be 3 10 cm in thickness. Scion sticks of the desired cultivar with 7 10 dormant buds are selected. Observe the progress of the branches of a top worked tree every 3 5 days. Cracks developed in the wax coating must be re-waxed. Whitewash the trunk to avoid sunburn. The scion must also be protected from the Sun by keeping the graft in shade. New shoots developing from scions are tied to stakes to avoid breaking off due to winds. New growth on older branches and trunk must be removed from time-to-time. The top worked trees must be regularly irrigated and manured.

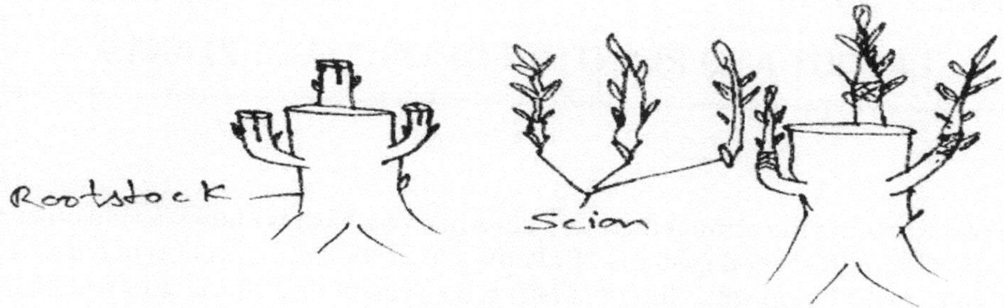

Top working

□□□

Exercise No. 5 **Date : ______________**

LAYOUT AND PLANTING OF ORCHARD PLANTS

The plan showing the arrangement of plants in an orchard is known as the orchard layout. Although several systems of planting are followed, selection of a suitable system depending on soil, climate, plant type, system of training and pruning is very important. Adoption of improper system results in over lapping of plant parts, competition for water, light, nutrient and unequal distribution of water etc, ultimately results in poor performance.

I- Square system:

This system is considered to be the simplest of all the system and is adopted widely. In this system, the plot is divided into squares and trees are planted at the four corners of the square, in straight rows running at right angles. While laying out the plot a base line is first drawn parallel to the road, fence or adjacent orchard, at a distance equal to half the spacing to be given between the trees. Pegs are fixed on this line at the desired distances. At both ends of the base line right angles are drawn by following the simple carpenter's 3, 4, 5 meters system. After the formation of three lines it is easy to fix all the other pegs to mark the tree locations in between the lines at the required spacing by using ropes connecting the pegs of the lines in opposite directions.

Under this system, intercultural operations, spraying, harvesting etc., can be done conveniently and easily. Planting of quick growing fruit trees like papaya, banana, guava during the early life of the orchard is possible. Raising of inter-crops like vegetables, ginger, turmeric, cumin, coriander and such other spices can be done conveniently. Cultivation and irrigation can be done in two directions.

Merits and demerits

The possibility of cultural operations in two directions is the greatest advantage of this system. The major disadvantage of this system is that a lot of space is wasted with in the squares.

Calculation of number of plants in square systems of Planting

$$\text{No. of plant} = \frac{\text{Area in metre}}{\text{Planting distance}}$$

Suppose area is 10000 sq. metre and planting distance is 10X10 metre, then

$$\text{Number of plant} = \frac{10,000}{10X10} = 100$$

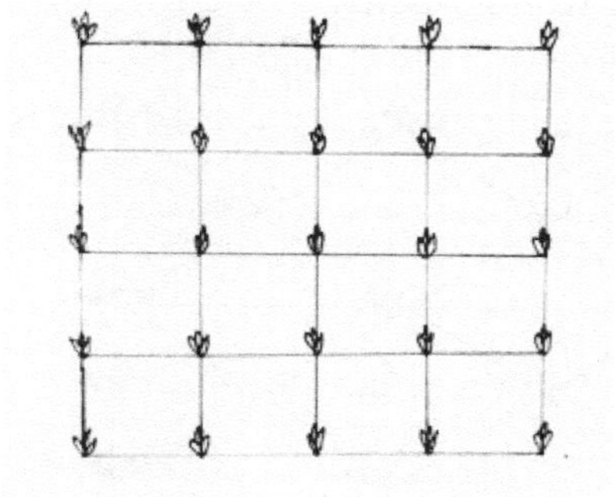

1. Squire System

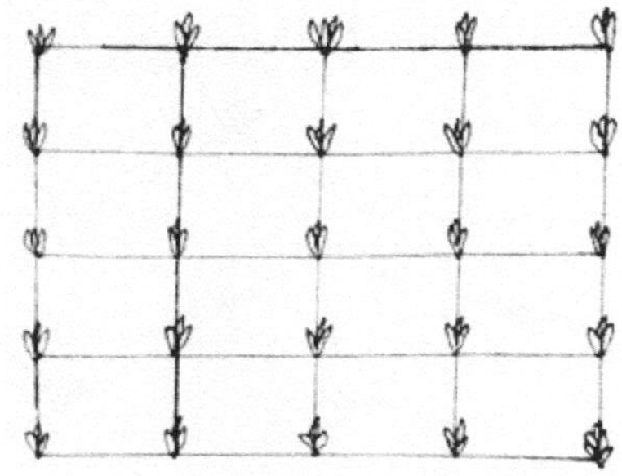

2. Rectangular System

II. Rectangular system:

In this system, the plot is divided into rectangles instead of squares and trees are planted at the four corners of the rectangle in straight rows running at right angles. The same advantages which have been mentioned in the square system are also enjoyed here. The only difference is that in this system more plants can be accommodated in the row keeping more space between the rows.

Merits and demerits:

This system is suited for such crops which require some support like trellis etc. e.g. grape and some ornamental creepers, which are trained to grow in two directions only. The major disadvantage in this system is that two way inter cultivation is not possible.

Calculation of number of plants in rectangular systems of Planting

$$\text{No. of plant} = \frac{\text{Area in metre}}{\text{Planting distance}}$$

e.g. Suppose , area in 10000 sq. metre and planting distance is 12X8 metre, then

$$\text{Number of plant} = \frac{10,000}{12X8} = 104.16 \text{ or } 104$$

III. Triangular system

In this system, trees are planted as in the square system but the plants in the 2nd , 4th, 6th and such other alternate rows are planted midway between the 1st, 3rd, 5th and such other alternate rows. This system has no special advantage over the square system except providing more open space for the trees and for intercrops. It is not only a difficult layout but cultivation also in the plots under this system becomes difficult.

Calculation of number of plants in triangular systems of Planting

$$\text{Number of plant} = \frac{\text{Area}}{(\text{Length of the triangle arm})^2} \text{X } 0.866$$

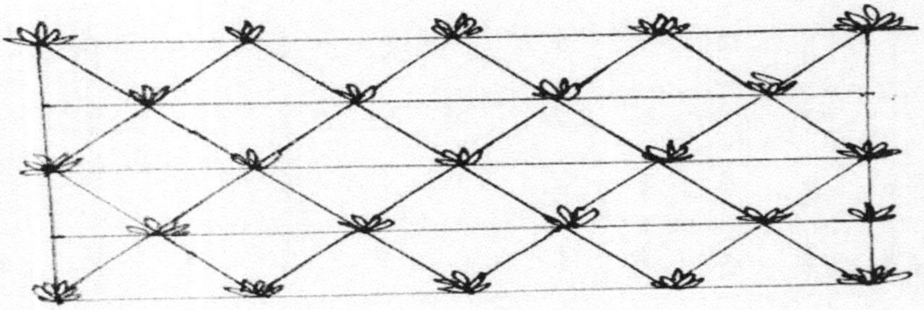

3. Triangular System

IV. Quincunx system:

This system of planting fruit trees is similar to square system, except that a fifth tree is planted at the centre of each square. As a result the tree number in an unit area becomes almost double the number in the square system. The additional tree in the centre is known as "filler". The fillers are usually quick growing, early maturing and erect type fruit trees like Phalsa, papaya, pomegranate, etc., which are removed as soon as the main fruit trees planted at the corner of the square come into bearing. The planting of filler trees provides an additional income to the grower in the early life of the orchard.

Merits & Demerits:

The plant population is about double than the square system. The difficulty in intercultural operations on account of the filler tree

Calculation of number of plants in quincunx systems of Planting

$$\text{No. of plants in squre system of planting} = \frac{\text{Area in metre}}{\text{Planting distance in metre}}$$

$$= \frac{10{,}000}{10\text{X}10} = 100$$

Additional plants planted in centres of squares = (No. of rows length wise-1) X (No. of rows width wise-1)

In 100X100 sq. metre field, if the planting distance is 10X 10 m, then No. of rows length wise and width wise will be 10

Hence, No. of plants = (10-1) X (10-1)

= 9 X 9 = 81

Total number of plants = Plants planted in square system of planting + additionally planted plants in the centres of squares

= 100 + 81 = 181

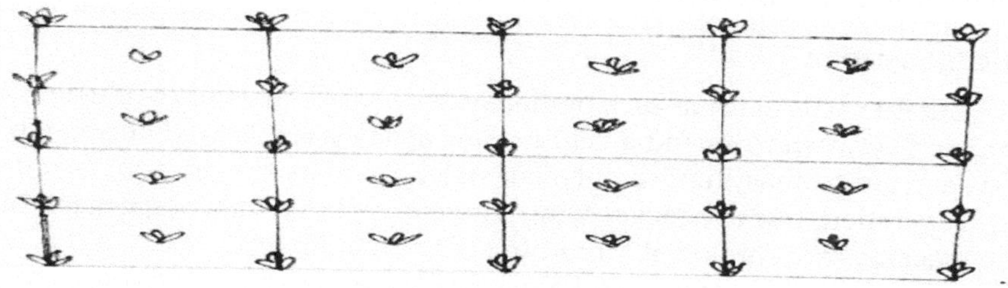

4- Quincunx system

V. Hexagonal system:

In this system, the trees are planted at the corners of an equilateral triangle and thus, six trees from a hexagon with the seventh tree at the centre. This system is generally followed where the land is costly and very fertile with ample provision of irrigation water. Though 15 per cent more trees can be planted in a unit area by this method over the square system, fruit growers usually do not adopt it, as it is difficult to layout and cultivation in the plot cannot be done so easily as in the square system.

Merits and demerits:

This system permits cultivation in three directions. The plants occupy the land fully without any waste as in square system. Beside this system allows planting of 15% more trees than can be planted by square system in a given area. But it is difficult to layout by this method and also the watch and ward becomes difficult as one cannot see in all the directions from a point.

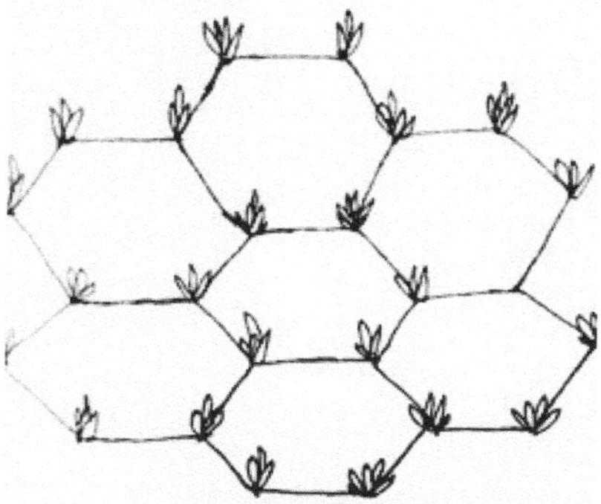

5. Hexagonal system

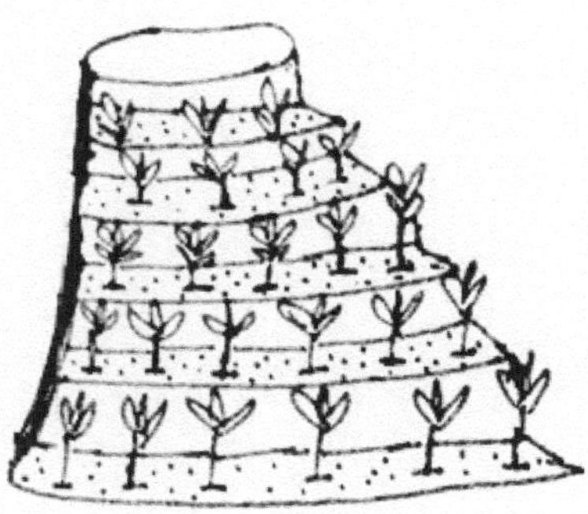

6. Contour system

VI. Contour system:

It is generally followed on the hills with high slopes. It particularly suits to a land with undulated topography, where there is greater danger of erosion and irrigation of the orchard is difficult. The main purpose of this system is to minimize land erosion and to conserve soil moisture so as to make the slope fit for growing fruits. So, the contour line is designed and graded in such a way that the flow of rain water becomes slow and thus finds time to penetrate into the soil without causing erosion.

VII. Hedge system:

The layout is exactly same as rectangular system except that very wider spacing is maintained between rows and a very narrow spacing is followed between plants. This system permits easy movement men, material and machinery and also effective cultural operations due to wider spacing. Therefore, this system is especially suitable where machines are employed for various farm operations.

7. Hedge system

❑❑❑

Exercise No. 6 **Date : ____________**

TRAINING AND PRUNING OF FRUIT TREES

Training:

Training is done primarily to train or shape the tree so that it has well distributed strong scaffold branches capable of bearing heavy crops over the years without breakage of limbs. Training starts from planting a tree. Fruit trees are trained so that the least amount of shading occurs inside the tree canopy. A tree that is pyramidal in shape exposes a greater percentage of total leaf surface to adequate sunlight.

Main points to be kept in view while training horticultural crops are;

(i) To admit adequate sunlight and to train the centre of the tree to expose more leaf area to the sun

(ii) To limit the growth and spread of the tree so that various cultural operations such as spraying and harvesting are performed at minimum cost.

(iii) To build the frame work and arrangement of scaffold branches.

(iv) To build the trees and maintain the height so as to reduce the exposure for sunscald and wind damage.

Systems Of Training:

1. Central leader:

In the central leader system of training the trunk is encouraged to form a central axis with branches distributed laterally up and down and around the stem. The central axis, or leader, is the dominant feature of the tree frame work and the main direction of growth is upward. This system of training is adopted in such types of trees which have a pronounced special dominance. Here the main trunk grows undisturbed. On account of vigorous (rapid) growth of the main trunk the tree develops a close centre and grows to great heights. The side branches remain more or less shaded and consequently they would be lower in vigour and productivity. Since the plants would be very tall the spraying and harvesting operations become difficult and costly e.g. pear and some varieties of apple. This is not encouraged now.

2. Open centre or vase shape:

In this system, the main stem is allowed to grow only up to certain height by heading it within a year of its planting or terminating its growth at a particular height and all the subsequent vegetative growth is promoted by lateral branches. However, special pruning is required to prevent a lateral from becoming dominant i.e. from forming a new central leader. The tree thus trained results in a low head and as such, the crop is borne closer to the ground in contrast to the central leader system. Open centre system allows the sunshine to be equally distributed to all the branches. The open centre trained trees are more fruitful besides greatly facilitating the operations like spraying, thinning and harvesting. However, the branches form narrow/weak crotches since they arise very close to one another almost from the same spot. So, there is certain amount of risk of splitting of the branches when there is a heavier load of crop on the trees. In areas of high light intensity, such trees suffer from sever sunscald and sun burn injuries. Example- peaches, Apricots, etc.

3. Modified Central leader:

This system stands intermediate between the central leader and the open centre, combined with the advantages of both the systems. It is developed first by training the tree to the leader type allowing the central stem to grow unhampered for first four to five years. The main branches are allowed to rise on the main stem at reasonable intervals. After the required number of them has developed the main stem is cut of. The top laterals will take the place of the main stem. This results in a fairly strong and moderately spreading type of trees, e.g., apple, walnut, mango, guava, citrus, jackfruit, bael and aonla.

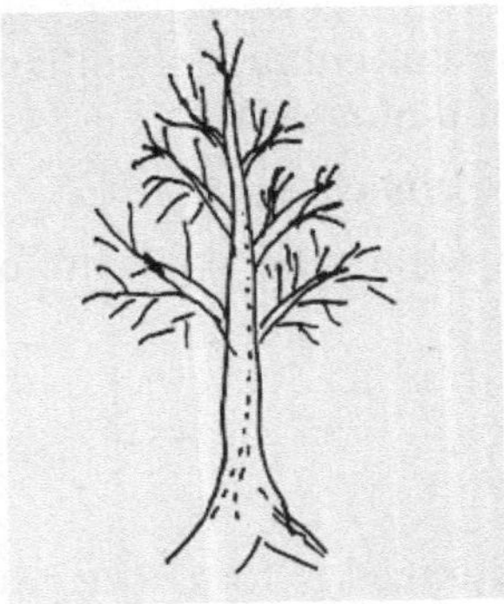

1. Central leader

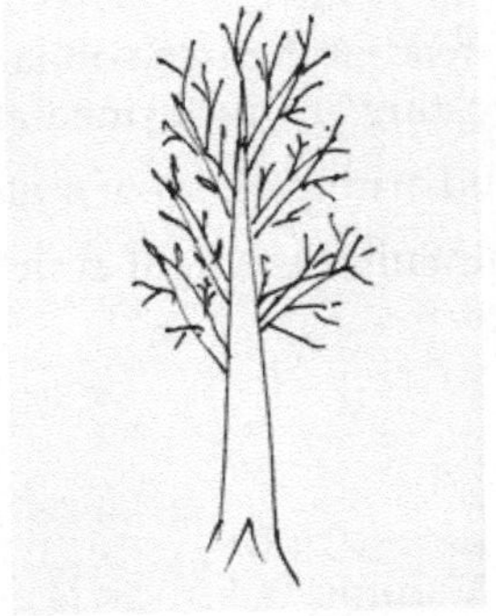

2. Modified Central leader

3. Open centre or vase shape

Other Methods Of Training:

1. Cordon system:

A single arm is allowed to develop from the trunk which will be trained along a stretched wire at a suitable height e.g., grape. An espalier restricted to one shoot or two shoots growing in opposite or parallel directions is called a Cordon system.

2. Kniffin System:

(a) Two arm kniffin system:

Two arms are allowed on the main stem at height of 5-6 ft and they trained on to wire on either side e.g., grape.

(b) Four arm kniffin system:

Here there will be four arms two each on either side of the trunk. The first pair of arms arise at a height of 3-4" and the second at 5-6" e.g., grape.

3. Overhead trellis or telephone system:

The wires will be fixed horizontally on some post just as telephone wires. The plant will be trained on to the wires with a suitable frame work, e.g., grape and ornamental creepers.

4. Pendal or Bower or Arbour system:

A pendal will be erected with the help of pillars and wires and the plant will be trained on to the pendal at 2.1 m height with suitable frame work, e.g., grape etc.

5. Single stake or umbrella systems or head system:

The main trunk will be supported by a stake. The trunk is beheaded at height of 5-6". The branches which arise on the trunk will be hanging freely. E.g.- grape, Var. Arkavathi.

Pruning Of Orchard Trees

Pruning is one of the major horticultural practices, which may be defined as "an art or a science of cutting away a portion of a plant". The parts more commonly removed are branches, leaves or both. Obviously, pruning is a subtraction process.

The extent and intensity of pruning on the same tree varies from year to year, depending on the growth of the tree, its bearing habit and season. The great majority of ornamental plants, if allowed to grow under normal conditions, become full sized trees growing in a haphazard way occupying a lot of space in the garden. When raised in pots to get the bushy growth at manageable sized and in planting proportion, one should prune them periodically, very few species seldom grow and remain dwarf and compact.

Main objectives of pruning:

1. To maintain the growth and vigour of the trees and to have a balance between the vegetative vigour and fruitfulness, so as to be conductive for production of optimum crop of best quality.
2. To shape the tree to make the best use of the space between trees while allowing the necessary access.
3. To regulate the size and quality of the fruits by way of proper distribution of the fruiting area.
4. To regulate the succession of crop and to have the crop where it can be managed easily and cheaply.
5. To spread the trees for economic orchard management.
6. To remove the dead, diseased and over aged wood.
7. For effective spraying of pesticides to the crop.
8. To minimize biannual bearing and consequent risk of die back.
9. To get maximum plageiotrophic shoots/stems.

Principles of pruning:

1. Excessive pruning should be avoided as it affects the growth of the plant by dwarfening and may induce more of water suckers, fasciations (union of a number of parts side by side in a flat plane) and thus affect the bearing potential.
2. In pruning, only that wood which is not necessary for the tree should be removed.
3. Pruning of larger limbs should be avoided as far as possible.
4. Pruning of young trees should be done more carefully than the yielding trees, since severe pruning of young trees delays the cropping and much more of yield area will be removed than what is desired.

Methods of pruning:

1. **Thinning out:** This refers to the removal of the branches entirely from its base leaving no stubs.
2. **Heading back:** This refers to pruning or cutting of main stem or all of few of the branches leaving a basal portion. This method is often followed for hedges, ornamental shrub and fore pruning in grapes.
3. **Disbudding or rubbing off:** Here the young buds are nipped without giving them the chance to sprout. The buds may be either vegetative or reproductive. This is practiced regularly in flowering plants to make the terminal bud to give a bigger flower. Thinning of the flower buds from the crowded one is essential to get large size quality blooms, as the lesser the number healthier and bigger the flowers. Generally, disbudding is done mainly for annuals, herbaceous perennials and roses. It is also followed in large flowered cultivars such as carnation, chrysanthemum, dahlia, marigold and zinnia. Generally one bud per shoot is retained.
4. **Pinching and topping:** This refers to the removal of the tip of the shoot alone with a view to stimulate mainly the lateral growth. This is practiced regularly in "coffee" to remove the apical dominance and to allow the side branches to grow vigorously.
5. **Time and extent of pruning:** The time of the year at which the different plants are pruned depends chiefly upon their dormant and flowering seasons. The best time for pruning most of the deciduous trees is at the end of the dormant season, i.e., about a month before the commencement of flowering, but time of pruning of evergreen tree should be during December- January when growth is at minimum as compare to dry and flowering seasons. The extent of pruning to be adopted for a particular crop depends on its growing and fruiting habit as it directly affects the nutritive condition within the tree and consequently affects the fruit formation. It also much depends on variety and age of the plants. Pruning of bearing and non-bearing trees differs. The non-bearing trees need much lesser pruning as compared to the bearing trees.

Exercise No. 7 **Date :** ____________

TRANSPLANTING AND CARE OF VEGETABLE SEEDLINGS

A. Transplanting

When seedling are ready for transplanting remove the shade a day before transplanting as this gives the seedling chance to get used to the strong sun-shine. Transplant late in the evening (3 to 6 pm).

B. Caring of vegetable seedlings

1. Watering

- The nursery beds require light irrigation with the help of rose can till the seeds get germinated.
- Excess rainwater or irrigated water should be drained out from the field as and when it is required otherwise plants may die due to excess of water.
- Watering in the beds depends upon the weather condition. If temperature is high, open irrigation is applied. Need not to irrigate the beds during rainy days.

2. Thinning

- It is an important operation to remove weak, unhealthy, diseased, insect pests damaged and dense plants from the nursery beds keeping distance of about 0.5 to 1.0 cm from plant to plant.
- The thinning facilitates balance light and air to each and every plant. It also helps in watching the diseased and insect pest attacked plants while moving around the nursery.

3. Weed control

- Timely weeding in nursery is very important to get healthy seedling. If there are some weeds in the seed bed, remove them manually either by hand or by hand hoe (thin forked Khurpi).
- Pre emergence herbicides can also be sprayed soon after seed sowing to control the weeds. Stomp @ 3 ml/litre of water should be sprayed on the nursery beds after the seed sowing and seed covering with mixture of FYM soil and sand.

4. Plant protection

Adaptation of plant protection measures in the nursery against the incidence of insect pest and diseases is very important task to get the healthy seedlings. Damping off seedlings, leaf curl, leaf blight diseases and leaf miner and borer infect the seedling in the nursery. The care for controlling them time to time is essential.

5. Damping off

- This is very serious disease of nursery.
- Post-emergence death of seeds is seen.
- In first instance rotting takes place on the stem near base of the stem and seedlings bent down near the ground and die.
- The causal organisms are pythium, phytopthora, rhizoctonia and Fusarium fungi.

Control

- Treat the nursery bed either by soil solarization, formalin solution or formalin dust or fungicides like thiram or captan as discussed earlier.

  The disease can be controlled to some extent by applying treated sand, soil and FYM mixture up to the level from where the seedlings are falling.

6. Leaf miner

It is very small sized insect enter in the leaves from margin side and move from one place to other by eating the leaf lamina. Initially the infected part of the leaves become brown and later on dry.

Control

- Spray 4% neem seed kernal extract on the plants (crush, add some water and allow them for overnight. In the morning filter the extract and makeup the volume 1000 ml.)
- Spray Monocrotophos or Metasystox 1.5 ml/litre of water.

7. Raising of virus free seedlings

Leaf curl is a white fly transmitted viral disease, infestation starts from seedling stage and continue till harvest of the crop. The disease is specially seen in the tomato and sometime in chilli too and causes great loss of the crop.

The leaves of affected plants show curling, mottling, rolling puckering etc.

Control

- Treat the soil of the nursery by carbofuran 3-5 g/sqm.
- Seed treatment with Imidachloprid @ 2.5 g/kg seed
- Cover the seed bed after seed sowing by Agronet making a tunnel like structure.

- Spray the nursery beds 15 days after seed germination at 7 days interval with Metasytox or Monocrotophos @ 1.5 ml/litre of water. Last spray should be done 2 days before transplanting.
- Removes the infected plants if any in the field and burried in with soil or burn.

8. Hardening of the plants in the nursery

- The term hardening includes

  "Any treatment that makes the tissues firm to withstand unfavorable environment like low temperature, high temperature and hot dry wind"
- Hardening is physiological process .Plants accumulate more carbohydrates reserves and produce additional cuticle on the leaves.
- In this process seedlings are given some artificial shocks at least 7-10 days before uprooting and transplanting. These shocks includes
- Exposure to the full sunlight,
- Removal of all the shedding nets, polythene sheets
- Irrigation should be stopped slowly and slowly.

9. Selection of seedlings for transplanting:

After attaining proper growth, seedlings are transplanted in main field. At the time of transplanting, seedling should be:

- Sticky and sturdy
- Should have good root system
- Should be free from any insect pests and diseases.

❏❏❏

Exercise No. 8 **Date : ____________**

MAKING OF HERBACEOUS AND SHRUBBERY BORDERS

Ornamental Herbs

A herbs is a short-sized plant with soft, green, delicate stem without the woody tissues. They complete their life cycle within one or two seasons. Generally, they have few branches or branchless.

Ornamental Shrubs

A shrub is a woody or semi-woody perennial plant with little or no trunk and grows up to a height from 50 cm to less than four meters.

Landscape uses of herbs and shrubs

Herbaceous borders

A herbaceous border is a collection of perennial herbaceous plants (plants that live for more than two years and are soft-stemmed and non-woody) arranged closely together, usually to create a dramatic effect through colour, shape or large scale.

Shrubbery borders

A shrubbery is a wide border to a garden where shrubs are thickly planted, or a similar larger area with a path winding through it. A single shrub is also known as a bush.

Hedge

When shrub is planted on boundary for protection/fencing, it is called as hedge.

Edge

When low growing plant are grown on the border of plot they are called as edge plants. They hardly grow upto 20-30 cm. A good edge plant is self supporting, tidy and compact. The important character of edge plant is reliable foliage that retains the colour and form of the plant.

Purposes of planting herbs and shrubs

- To enhance the beauty of surroundings
- To provide fragrance in the garden
- To act as boundary of property line

- To secure privacy for the family
- To divide different areas or features in the garden
- Softening harsh building lines
- To screen off unwanted sites/ features
- To reduce winds
- To control traffic
- To provide background for growing annual flowers
- To reduce soil erosion
- To conserve moisture
- To discourage weed growth
- To provide natural habitat for wildlife especially birds

Basis of shrubs for arranging in the shrubbery border:

- Height
- Flower colour
- Flowering season
- Foliage colour
- Leaf retention time

Types of shrubbery borders

- Single faced: Tall- Medium- Dwarf. Created on the corner of the garden.
- Double faced: Dwarf- Medium-Tall- MediumDwarf. Created in the middle of the garden.

Selection of shrubs

1. **Shrubs valued for ornamental fruits**
    - *Ochna squamosa*
    - *Duranta plumier*
    - *Solanum pseudocapsicum*
    - *Euonymous japonicas*
    - *Citrus japonica*
2. **Shrubs valued for ornamental coloured foliage**
    - *Acalypha hispida*
    - *Codiaeum variegatum*
    - *Excoecaria bicolor*
    - *Euphorbia cotinifolia*
    - *Gynura bicolor*

3. **Shrubs valued for ornamental flowers and foliage**
   - *Camellia japonica*
   - *Hamelia patens*
   - *Hibiscus rosa-sinensis*
   - *Gardenia jasminoides*
   - *Nerium indicum*
4. **Shrubs valued for fragrant flowers**
   - *Jasminum species*
   - *Gardenia jasminoides*
   - *Cestrum nocturnum*
   - *Camellia japonica*
   - *Murraya exotica*
5. **Shrubs valued for ornamental bracts**
   - *Euphorbia pulcherrima*
   - *Mussaenda frondosa*
   - *Mussaenda philipica*
   - *Bougainvillea species*
   - *Flemingia fruticulosa*
6. **Shrubs flowering almost through out year**
   - *Beloperone amherstiae*
   - *Cassia laevigata*
   - *Euphorbia milli*
   - *Justicia magnifica*
   - *Vinca rosea*

❑❑❑

Exercise No. 9 **Date : ___________**

PREPARATION OF POTTING MIXTURE, POTTING AND REPOTTING

Plants should be moved into larger containers as they grow. Unless more space is provided for the plant's roots, they can become pot-bound. That is, the roots of the plant become cramped and form a tightly packed mass that inhibits growth.

How do you know if a plant needs repotted?

1. The most obvious sign is when you can see roots on the surface of the soil or emerging from the drainage hole in the bottom of the pot.
2. If the plant seems to stop growing or has slowed growth, it has likely become pot-bound. If it's a small plant, turn the pot on its side and ease the plant out of its container. Take a look at its roots.

   Are they coiled in the bottom of the pot? If so, it's definitely time to repot.
3. Offsets produced by plants can become crowded in the pot and need to be separated and propagated in their own containers.

When to repot a plant

1. If your plant just came home from the garden center, let it adjust to its new environment for a couple of weeks before repotting it. Plants are in shock until they get used to new light, temperature, and humidity conditions. If you want to cover up a plain plastic container, put it in a cachepot.
2. Young, actively growing plants should to be moved into slightly larger pots with fresh potting mix once a year. Repotting house plants that are large, such as ficus, or slow-growing plants can be done every two years or when they seem to outgrow their pots or look top-heavy. If a plant is thriving, you can assume it is happy in its pot.
3. It's a good idea to repot a plant at the beginning of a period of active growth, usually in spring. Repotting house plants that bloom in winter should be done in early fall, after their dormant period.

Pots of different sizes

How to choose a container

1. The new pot should not be more than 2 inches wider at the rim -- or 2 inches deeper -- than the old pot. A much larger pot gives the too much space for the

roots to grow. The top of the plant won't grow until it roots begin to fill the container. A too-large container will also hold too much water and can cause root rot. Also, while choosing a pot, be sure that the container has a drainage holes to allow excess water to escape.

2. Scrub used pots between plantings to remove any diseases. You can disinfect a pot by soaking it in a solution of one part chlorine bleach to nine parts water. Rinse well with clear water. If you're using a new terra cotta pot, soak it in water for a few hours before you plant in it. New terra cotta is so dry that it will rob moisture from the soil leaving the plant thirsty.

How to repot a plant

To remove the plant from its original pot, turn it on its side and ease the plant gently from the pot. If the plant won't budge, you may have to tap the bottom of the pot on a hard surface to loosen it or, slide a trowel or knife around the inside of the pot, taking care not to damage the rootball.

1. If the roots are coiled around the bottom, use your fingers to pull them straight. Prune the roots before repotting. Pruning will stimulate new root growth and help the plant to establish in its new container.
2. Partly fill the new container with potting mix. Center the plant in its container, then fill the sides of the plant with additional mix. Tamp it down with your fingers, especially around the sides of the pot.
3. Water thoroughly to moisten roots and to settle potting mix. Add more mix if needed.

After-repotting house plants care Tips

1. Repotting house plants is stressful for them and they need time to recuperate. Here are a few tips to help
2. Plants adjust: Do not expose it to direct sun right away because sun can be too harsh on a weakened plant.
3. Keep the soil evenly moist, but not soggy. If you notice that the leaves are limp, the plant is not getting enough water. If the edges of the leaves turn brown, it's getting too much water.
4. High humidity sometimes helps a newly repotted plant recover.
5. Never fertilize a newly repotted plant. Its roots have likely been cut and can suffer from fertilizer burn. Wait at least a month before fertilizing when its root system is better established.

Potting mixtures, containers and repotting

A loamy soil with a mixture of organic and mineral materials (50% space and 50% matter), to ensure optimum plant growth and necessary nutrition. The main ingredients in potting soil are sphagnum peat moss, vermiculite or perlite and aged compost products. Sphagnum peat moss is responsible for holding water in the soil. Perlite separates the fibers in the peat moss so the soil is more porous. Vermiculite has the same function but holds more water than perlite. Compost adds nutrients to the soil and the best quality compost is made from aged forest products. Potting mixes are sometimes referred to as "soilless" meaning that they do not contain the type of dirt you would find in your yard or garden. The best soil for most house plants contains 2 parts loam, 1 part composted organic matter and 1 part coarse, clean sand plus 1 tablespoon powdered phosphate rock and granite dust (greensand or potash) each to each mixture for each pot. This replaces the need to use chemicals or manure tea in potted plants.

Charcoal added to either mix keeps the soil sweet and improves drainage. Scoop up a handful of the soil mix and squeeze. If it holds together until touching causes it to crumble easily, you know you have the right amount of moisture in the soil. When reusing soil, you must first pasteurize it in a tray in a heated oven at 180° F for 30 minutes. Allow to cool before use. If pasteurized soil becomes too dry when heated, place in a plastic bag, add enough water to moisten the soil. Do not soak. Then try the hand test again. Plastic pots are lighter in weight than clay pots. However, the clay pots allow the soil and roots to breathe. Water evaporates easier in a clay pot. If you use plastic containers for your plants, you will need to water and feed less. Make certain all your pots have drainage holes in the bottoms. Because these pots will be used in your home, always use plastic or glass saucers and pads to prevent moisture damage to your furniture.

Exercise No. 10 **Date :** ____________

MANURE AND FERTILIZER APPLICATION IN DIFFERENT CROPS

The tern "**manure**" was used originally for denoting materials like cattle manure and other bulky natural substances that were applied to land with an objective of increasing the production of crops. Later, chemical substances, ammonium sulphate and super phosphate containing plant nutrients in abundance were used for increasing crop production and were also called manures. So in a general way, manure is a substance containing plant nutrients, which is applied to the soil for increasing crop production. To distinguish organic and inorganic substances used for the purpose, the organics were later specially called manures and inorganic were called fertilizers. This usage is commonly accepted, though the term manure is oftenly used to refer to organic manures only and both organics and in-organics in a general way.

Artificial fertilizers:

They are inorganic substances which supply plant nutrients, though in recent times some organic substances like Urea have also been included in their category. They are rich in plant nutrients and supply individually one or two or three elements of plant food only.

Manure

I. The organic manures:

They are made up of dead plant and animal remains. They are generally bulky in nature and poor plant food. They supply almost all the ingredients of plant food and are therefore general manures.

A. Bulky organic manures:

(i) Farm yard manure (FYM):

It is also known as stable manure, barn manure, dung and cattle manure. It is made up of excreta of farm animals, the litter or bedding provided for them and miscellaneous farm and house hold wastes. Nutrient availability in FYM is 0.5% N, 0.4-0.8% P2O5 and 0.5 1.9%, K2O.

(ii) Compost:

All organic farm wastes, stubbles of crops, waste straw, sweepings, threshing floor collections, dry weeds etc. may be collected and used for augmenting the organic matter supplies. They cannot be used directly as manure in most cases and required to be decomposed before application. The process of decomposing organic wastes is called composting and the decomposed material is called "compost". It does not contain plant food as much as cattle manure. Since the value given to plant food is much as cattle manure. Since the value given to cattle manure is more for the organic matter than for the plant food, composts are also equally valuable. The organic matter in composts is also equally valuable. The organic matters in compost and cattle manure have the same properties and value. Compost (Urban): 1.0-2.0% N, 1.0% P2O5 and 1.5% K2O. Compost (Rural): 0.4-0.8% N, 0.3-0.6% P2O5 and 0.7-1.00% K2O.

(iii) Green manuring and green leaf manuring:

These are important practices followed to supply organic matter collecting green leaves from all available sources and using it for manuring wet land is an age old practice in South India which is called green leaf manuring. Growing a crop purposely and incorporating it in the soil for manuring is called green manuring. Preferably legumes alone are used as green manure crops. Green manures contains (various averages); 0.5 0.7%N, 0.1-0.2% P2O5 and 0.6-0.8% K2O.

Examples:

(a) Green manures: Jute, Crotalaria juncea (sunhemp), Diancha, Sesbania speciosa.

(b) Green leaf manures: Cassia auriculata, Dodonia viscoa, Pongamia glabra, Cassia siamea, Gliricidia macculata.

B. Concentrated organic manures:

(i) Manures of plant origin:

		Nutrient content (%)		
S.No.	**Manures**	**N**	**P_2O_5**	**K2O**
a.	Castor Cake	5.5 – 5.8	1.8 – 1.9	1.0 – 1.1
b.	Neem Cake	5.2 – 5.3	1.0 – 1.1	1.4 – 1.5
c.	Safflower cake	4.8 – 4.9	1.4 – 1.5	1.2 – 1.3
d.	Coconut	3.0 – 3.2	1.8 – 1.9	1.7 – 1.8
e.	Groundnut	7.0 – 7.2	1.5 – 1.6	1.3 – 1.4
f.	Niger	4.7 – 4.8	1.8 – 1.9	1.3 – 1.4
g.	Sesamum cake	6.2 – 6.3	2.0 – 2.1	1.2 – 1.3

(ii) Manures of plant origin:

a.	Fish mean, fish manuring and fish guano	4.0 – 10.0	3.0 – 9.0	0.3 – 1.5
b.	Bone meal (Raw)	3.0 – 4.0	20.0 – 25.0	-

c.	Bone meal (Steamed)	1.0 – 2.0	25.0 – 30.0	-
d.	Settled sludge (Dry)	2.0 – 2.5	1.0 – 1.2	0.4 – 0.5
e.	Night soil	1.2 – 1.3	0.8 – 1.0	0.4 – 0.5
f.	Human urine	1.0 – 1.2	0.1 – 0.2	0.2 – 0.3
g.	Cattle dung and urine mixed	0.60	0.15	0.45
h.	Horse dung and urine mixed	0.70	0.25	0.55
i.	Sheep dung and urine mixed	0.95	0.35	1.00

Fertilizer:

The inorganic simple fertilizers supplying Nitrogen, Phosphorus and Potassium.

(i) Nitrogen fertilizers:

S.No.	Fertilizer Name	Main nutrient (%)		Other nutrients (%)	
		N	Na	S	Ca
a.	Urea	46	-	-	-
b.	Calcium nitrate	15.5	-	-	20
c.	Ammonium sulphate	21	-	24	-

(ii) Phosphorus fertilizers:

S.No.	Fertilizer Name	P_2O_5	P	-	-	
a.	Super phosphate	18-20	8-9	-	12-14	20
b.	Rock phosphate	29-36	12.5-15.5	-	-	31-35
c.	Single super phosphate	16	7	-	-	-
d.	Diammonium phosphate	46	-	P-18	-	-

(iii) Potassium fertilizers:

S.No.	Fertilizer Name	K_2O	K	-	-	-
a.	Muriate of potash	60	50	-	-	-
b.	Sulphate of potash	50	42	-	17	-

Bio fertilizers:

A. Symbiotic N fixing system:

(a) Rhizobium: 50-500 kg/ha/season depending upon the leguminous app.

B. Free lying N fixing system:

I. Aerobic bacteria; like Azotobacter, Beijerinckia, Klebsiella.

II. Anaerobic bacteria; Clostridium

(b) Photosynthetic: Anabaena, Nostoc, Aulosira, Tylopothrix etc. Ex. Blue green algae and Azotobacter – paddy.

C. Associate symbiotic system:

(a) Azospirillum lipoferum

(b) Phosphobacteria free living; Bacillus megaterium: B. circullaus convert insoluble P to soluble P.

Methods of application of manures and fertilizers:

Manures:

Usually FYM/compost is applied at the time of land preparation at least 15 days before sowing or planting in respect of annuals or short duration crops. However, in the case of perennials, they are applied to pits at the time of planting and to basin during subsequent years. Similarly, if a green manure crop is grown it will be ploughed back to the soil before flowering and well before the crop is sown/ planted to ensure proper decomposition and availability of nutrients for early growth and development.

Fertilizers:

Various methods of application of fertilizers are in vogue and following are few important practices widely adopted.

1. **Broadcasting :** Broadcasting on the soil surface after ploughing and mixing with the surface soil by harrowing.
2. **Row application :** Applying fertilizers in bands, 5 to 8 or more centimetres from the row and 5 to 8 or more centimetres below the surface.
3. **Side placement :** Acombination of broadcasting methods or plough furrow application with bands at the side of the row to planting time.
4. **Perforated:** Applying fertilizer with a drill below the surface of the soil before crop is planted.
5. **Liquid :**

   (a) By direct application to the soil.

   (b) Foliage spray

   (c) Irrigation water.

Time of application of sufficient moisture is very: Depending on the crop, annual, biennial or perennial or variety time of application varies. The time of application also varies according to type of nutrient one would like to supply. Usually P and K are applied basally at the time of sowing or planting. But nitrogen is applied in varying number of splits depending on the crop and its duration. In any case fertilizers should be made available at right time in right quantity. Availability is important.

Exercise No. 11 **Date : ____________**

VISITS TO COMMERCIAL NURSERIES/ORCHARD

Introduction

Nursery is consequently the basic need of horticulture. Plant propagation techniques and practices is the core of horticulture nurseries. The planting materials for horticultural plantations are raised from seeds and vegetative parts. Role of Mother Plants is very primary and important. The fate of nursery depends on quality and truthfulness of mother plants. A good nursery entrepreneur does not depend on others for procurement of mother plants. Mother plants are required for both stock and scion. Mother plants should be selected on the basis of its genetic traits and other factors like availability and adaptation in the growing environment.

Types of Nurseries

- According to type of plant grown
- According to type of demand

Types of Nurseries According to Type of Plants Grown

1. **Fruit plant Nurseries**

   Fruit crops are mainly propagated vegetatively and need special techniques for propagations as well as maintenance. Mango, Guava, Pomegranate, Sapota, Oranges etc. are propagated with vegetative means. Fruit nurseries are essential for production of grafts as well as the mother plants of scions and rootstocks.

2. **Vegetable Nurseries**

   All vegetables except few like potatoes, sweet potato, bulbous vegetables and some other are raised by seedlings. Very few vegetables are perennials like, little gourd, drumsticks, Alocasia etc. Seedlings are to be produced on a large scale in short period.

3. **Ornamental Plant Nurseries**

   Ornamental and floricultural crops are numerous and are propagated vegetatively, like gladiolus, carnation, roses, lilies etc. There is a large group of ornamental plants, which is propagated by seeds and seedling; Asters, Marigolds, Salvias, etc. are some of them.

4. **Medicinal and Aromatic Plant Nurseries**

There is considerable increase in people adopting ayurvedic medicines with the changing life style. It is also necessary to conserve the fast depleting precious medicinal and aromatic plants. To save and multiply the valuable medicinal and other auspicious plants, nurseries specializing in these plants have begun to flourish. These plants are in grate demand by the Ayurvedic medicinal practitioners.

5. **Hi-Tech Nurseries**

There is sudden increase in the demand for certain commercial plants. For example Tissue cultured banana, gerbera and carnation etc. It is not possible to fulfill this requirement by ordinary or common nursery practices. There is necessity to have special techniques and methods to meet the demand and only Hi-tech nurseries can satisfy this type of demand. These nurseries grow plants in greenhouse, building of glass or a plastic tunnel, designed to protect young plants from harsh weather, while allowing access to light and ventilation. Modern greenhouses allow automated control of temperature, ventilation, light, watering and feeding. Some also have fold-back roofs to allow "hardening-off" of plants without the need for manual transfer of plants to the outdoor beds.

Types of Nurseries According to the Type of Sale

1. **Retail Nurseries:**

Retail nurseries raise plants for sale to the general public. These places are small, locally owned nurseries that sell seasonal, annuals, ornamental trees, other landscaping plants and garden decoration to the general public or companies that specialize in a particular type of plant, such as tropical plants, citrus trees, bulbs or roses.

2. **Wholesale Nurseries:**

Wholesale nurseries usually grow plants in bulk for the purpose of selling to large clients. These clients may include florists, garden centers or departmental stores. A wholesale nursery may fill a niche for particular types of plants, such as vegetables or houseplants, or they may grow a general selection of plants to sell such as fruits, vegetables and landscaping plants.

3. **Private Nurseries:**

A private nursery grows plants exclusively for a single client. The private nursery may be owned by the client or it may be under contract for use by the client. Clients for private nurseries include large estates, corporations and institutions. These nurseries are concerned with raising documented historical plants for the historic preservation of the estates.

4. Mail Order:

Privately owned, retail and wholesale businesses may all be involved in mail order businesses. As shipping technology improved, it became possible to ship dormant ornamental trees and bedding plants via mail. The internet has largely shifted mail order from catalog to online shopping. Bedding plants may be shipped via postal carrier, but are primarily handled through third-party shipping agents.

❑❑❑

Exercise No. 12 **Date : ___________**

NURSERY ACTS

Introduction

In India, more than 4409 fruit nurseries including, 1575 Government and 2834 private nurseries are functioning which, are having annual target, to produce quality seeds and planting materials accounting to 1387 million plants and 2468 million seeds of fruits, Vegetables, spices, coconut, cashewnut, arecanut, cocoa and medicinal plants. In order to maintain the quality of planting materials, Govt. of India, Ministry of Agriculture, has passed the fruit nurseries regulation acts in each state to function the registered nurseries under certain acts of licensing and regulation of fruit nurseries.

The Uttar Pradesh Fruit Nurseries (Regulation) Act, 1976

The commencement of the Uttar Pradesh Fruit Nurseries (Regulation) Act came into force in Uttar Pradesh w.e.f July 1, 1976 Vide Note-No-3074/12(4)-173-75 dated June 30, 1976 and Gazetted as Notice No-6275/XII-4-173-75 dated 15 October, 1976 pp-8-15.

(A) Definitions of Act:

The Act/Bill is introduced to provide licence and regulation of fruit nurseries. The definition of in the act is as below:

1. **Fruit nursery:** Means any place where fruit plants are in the regular and course of business, propagated and sold for transplantation. The fruit nursery should have more than 0.2 hectares and managed by private sector/entrepreneurs.
2. **Fruit plant:** Means any plant, which can produce edible fruits, nuts, and includes bud wood, seedlings, grafts, layers, gootee, seeds, suckers, rhizomes and cuttings.
3. **Licence:** Means a licence granted by competent authority of State Govt.
4. **Licensee:** Means any person holding a licence.
5. **Licensing authority:** Means the Director of Horticulture and Food Processing, U.P. Govt. and includes not below the rank of District Horticulture officer, empowered by Director on his behalf.
6. **Owner:** Means the person who or the authority which a scion variety is grafted / budded.

7. **Rootstock:** Means a fruit plant or part on which a scion variety is grafted / budded.
8. **Scion:** Means a portion of fruit plant, which is grafted / budded on the rootstocks.

B. Application of Licence:

The application for grant of licence under this Act shall be applied on prescribed form No. 1.

C. Renewal of Licence:

The licence is renewed after expiry of valid period prescribed in granted licence (Form No. 2) and it is applied in prescribed form No. 3.

D. Duplicate licence:

If any licence lost, damaged, defacted, torn, illegible, the licensing authority may issue duplicate licence on prescribe form No. 4 along with charging of payment fee.

E. Duties of the licence:

1. Employ only such varieties of the fruit plants specified in the licence in respect to stock-scion.
2. Keep a complete record of source of rootstock and scion including, common name, botanical name and variety/clone.
3. Keep the nursery plots as well as the progency mother plants clean.
4. Production of fruit plants as directed by 'Licencing Authority'.
5. Specify in a conspicuous manner on label the name of every variety, name of nursery and age of graft.
6. Sale or distribution of any plants completely free from any insect- pest and diseases.
7. Licencee maintained account 'records/books/register and annual balanced sheet of nursery production.
   (a) Cash receipt book (Form No. 7)
   (b) Stock Register (Form No. 8)
   (c) Fruit plant propagation register (Form No. 9)

F. Powers of licensing authority:

1. Issue and cancelling of nursery licence under specific terms and conditions prescribed in the Acts.
2. Renewal of licence.
3. Prohibition and regulation of sale, import, export or transportation of certain plants.

4. Punishment on any offences and violation of Acts. Penality and conduct of business as per terms and conditions of licence.
5. Inspection of nursery records and account book.
6. Certification and testing of plant samplings.
7. Hearing of appeals of licencee for any grievances, cancellation, renewal, suspension and refusal of licence through Appeliate Authority of State Govt. under sub-section (a) of Section 6.

❑❑❑

Appendix-1

HORTICULTURE SCENARIO (INDIA)

1. Horticulture Statistics: 2018-19

- The percentage share of horticulture output in Agriculture has become 33%
- Under the preview of Agriculture & allied activities, the share of plan outlay for Horticulture which was 3.9% during 9th Plan, has increased to 4.6% during the 12th Plan.
- Note: State having maximum area and production under Total Horticulture i.e. U.P.

2. Per capita availability of Fruits and Vegetables in India

- Projected Population of India in 2017-18 (million) - **1283**
- Fruits Production in 2017-18 (thousand tonnes) - **97358.0**
- Vegetables roduction in 2017-18 (thousand tonnes) - **184394.0**
- "Per Capita Availability of fruits (g/person/day)" - **207.90**
- "Per Capita Availability of vegetables (g/person/day)" - **393.76**

Origin of Horticultural Science

- The origin of horticultural science derives from a coming together of three events.
- The formation of scientific societies in the 17th **century**.
- The creation of agricultural and horticultural societies in the 18th **century**.
- The establishment of state- supported agricultural research in the 19th **century** in different countries.
- In England two seminal horticultural societies involved.
- The Horticultural Society of London (later the Royal Horticulture Society) founded in **1804**.
- The Society for Horticultural Science (later the American Society for Horticultural Science) **ASHS** funded in **1903, Alexandria USA**.

3. Production of Horticulture Crops (2018-19)

- Total Horticulture production in India is **314.67 million tonnes**

- Total Horticulture area in India is **25.87 million hectare**
- Share of Horticulture sector in agriculture **GDP is 30.4**
- India is second largest producer of fruit and vegetable crops **(1st China)**

4. Fruit Production

- Total Production of fruit crops is **97.35 million tonnes**
- Total Area under fruit crops is **6.51 million hectare**
- The Productivity of fruit crops is **14.3 MT per hectare**
- India accounts 3% of total world production of fruits
- India is a second largest producer of fruits in the world
- India ranks first in production of **Mango, Banana, Papaya, Sapota, Pomegranate, Acid lime and Aonla in the world**
- India has highest productivity in **grape in the world**
- State with highest fruit production is **Andhra Pradesh (13%), Maharashtra (11.2 %)** and **U.P (11.2%)**
- Production wise fruit in India is **Banana (33%), mango (21%), citrus (14%)**
- Area wise fruit in India is Mango, Banana, Citrus
- Fruits accounts for about **11%** of total horticultural export from country.
- Grape and mango together constitute **60%** of India's exports of fresh fruits

Total Area, Production and Productivity of different Horticultural Crops (2018-19)

Crops/ Plants	Area (Mha)	Production (MT)	Productivity (Mt/ha)
Fruits	6.53	97.35	14.96
Vegetables	10.43	187.47	17.97
Cut flower	0.39	0.86	-
Loose flowers	0.39	1.99	-
Total flower	0.39	2.85	3.48
Plantation crops	3.76	18.08	4.59
Spices	4.08	8.50	2.00
Medicinal & Aromatic Crops	0.72	0.89	-
Total Horticulture Crops	**25.87**	**314.67**	**12.25**

Appendix-2

AGRICULTURE INSTITUTION

1. **Acharya N G Ranga Agricultural University (ANGRAU)** 1965
   Rajendranagar, Hyderabad, Andhra Pradesh PIN 500030

2. **Anand Agricultural University (AAU)** 2003
   Anand, Gujarat

3. **Assam Agriculture University (AAU)** 1969
   Jorhat, Assam PIN 785013

4. **Bidhan Chandra Krishi Vishva Vidyalaya (BCKVV)** 1974
   Haringhatta PO, Mohanpur, Nadia, West Bengal PIN 741246

5. **Birsa Agricultural University (BAU)** 1980
   Kanke, Ranchi, Jharkhand PIN 834006

6. **Central Agricultural University (CAU)** 1995
   Imphal, Manipur PIN 795001

7. **Ch. Sarwan Kumar Krishi Vishwa Vidyalaya (CSKHPKV)** 1978
   Palampur, Himachal Pradesh, PIN 176062

8. **Chandra Shekhar Azad University of Agriculrure & Technology (CSAUT)**
   Kanpur, Uttar Pradesh PIN 208002 1975

9. **Ch. Charan Singh Haryana Agricultural University (HAU)** 1970
   Hisar, Haryana PIN 125004

10. **Dr. Balasaheb Sawant Konkan Krishi Vidyapeeth (KKV)** 1972
    Dapoli, Maharashtra PIN 415712

11. **Dr. Panjabrao Deshmukh Krishi Vishwa Vidyalaya (PKV)** 1969
    Krishi Nagar, Akola, Maharashtra PIN 444104

12. **Dr. Yashwant Singh Parmar University of Horticulture & Forestry (YSPUH&F)**
Solan, **Himachal Pradesh** PIN 173230 1985

13. **Govind Ballabh Pant University of Agriculture and Technology (GBPAU&T)**
Pantnagar, **Uttaranchal** PIN 263145 1960

14. **Indira Gandhi Krishi Vishwa Vidyalaya (IGKVV)** 1987
Raipur, **Chhattisgarh** PIN 492012

15. **Jawaharlal Nehru Krishi Vishwa Vidyalaya (JNKVV)** 1964
Jabalpur, **Madhya Pradesh** PIN 482004

16. **Junagadh Agricultural University (JAU)** 2003
Junagadh, **Gujarat** PIN 362001

17. **Kerala Agricultural University (KAU)** 1972
Vellanikkara, Trichur **Kerala** PIN 680654

18. **Maharana Pratap University of Agriculture & Technology (MPUAT)** 1986
Udaipur, **Rajasthan** PIN 313001

19. **Maharashtra Animal Science & Fisheries Sciences University(MASFSU)**
Nagpur, **Maharashtra** PIN 440006

20. **Mahatma Phule Krishi Vidyapeeth (MPKV),** Rahuri, **Maharashtra** 1965

21. **Marathwada Agricultural University (MAU)** 1972
Parbhani, **Maharashtra** PIN 431402

22. **Acharya Narendra Deva University of Agriculture and Technology (ANDUAT) Ayodhya, Uttar Pradesh** PIN- 224229 1976

23. **Navsari Agricultural University (NAU)** 2003
Navsari, **Gujarat**

24. **Orissa University Of Agiculture & Technology (OUAT)** 1962
Bhubaneswar, **Orissa** PIN 751001

25. **Punjab Agricultural University (PAU)** 1962
Ludhiana, **Punjab** PIN 141004

26. **Rajasthan Agricultural University (RAU)** 1986
Bikaner, **Rajasthan** PIN 334002

27. **Rajendra Agricultural University (RAU)** 1971
Pusa, Samastipur, **Bihar** PIN 848125

28. **Sardar Vallabh bhai Patel University of Agriculture & Technology (SVBPUAT)**
Modipuram, Meerut, **Uttar Pradesh** PIN 250110 1996

29. **SardarKrushiNagar-Dantiwada Agricultural University (SDAU)** 1972
Sardar Krishi Nagar, Banaskantha **Gujarat** PIN 385506

30. **Sher-e-Kashmir University of Agricultural Sciences & Technology (SKUAS&T, Jammu)**
Camp Office, Railway Road, **Jammu & Kashmir** PIN 180004 1998

31. **Sher-e-Kashmir University of Agricultural Sciences & Technology (SKUAS&T, Kashmir)**
Srinagar, Shalimar Campus, Srinagar, **Kashmir** PIN 191121 1982

32. **Tamil Nadu Agricultural University (TNAU)** 1972
Coimbatore, **Tamil Nadu** PIN 641003

33. **Tamil Nadu Veterinary & Animal Sciences University (TNU&ASU)** 1989
Chennai, **Tamil Nadu** PIN 600007

34. **University of Agricultural Sciences (UAS)** 1965
Bangalore, **Karnataka** PIN 560065

35. **University of Agricultural Sciences (UAS)** 1987
Krishi Nagar, Dharwad, **Karnataka** PIN 580005

36. **UP Pandit Deen Dayal Upadhyay Pashu chikitsa Vigyan Vishwavidyalaya evam Go**
Anusandhan Sansthan, Mathura, **Uttar Pradesh PIN 281001**

37. **Uttar Banga Krishi Vishwavidyalaya (UBKV)**
Distt. Cooch Behar, **West Bengal** PIN 736165

38. **West Bengal University of Animal & Fishery Sciences (WBUA&FS)**
Sarani, Belgachia, Kolkata, **West Bengal** PIN 700037

Appendix-3

FAMOUS NAME OF HORTICULTURAL CROPS

S. No.	Famous Name	Crops
1.	King of fruit	Mango
2.	King of arid and semi arid fruits	Ber
3.	King of Temperate Fruit	Apple
4.	Queen of fruits	Litchi, Mangosteen
5.	King of spices	Black Pepper
6.	Queen of spices	Cardamom
7.	King of Vegetabes	Potato
8.	Queen of Vegetable	Asparagus
9.	Poor man's fruit	Ber
10.	Poor man's friend	Potato
11.	National fruit of India	Mango
12.	Glory of East	Chrysanthemum
13.	Autumn queen	Chrysanthemum
14.	Queen of Night	Cestrum nocturnum
15.	King of flower	Rose
16.	Queen of flower	Jasmine
17.	Egg plant	Brinjal
18.	Butter fruit	Avocado
19.	Queen of beverage crop	Tea
20.	China's miracle fruit	Kiwi fruit
21.	Food of God	Cocoa
22.	Small holder's irrigated crops	Oil palm
23.	Oldest cultivated tropical fruit	Banana
24.	Tree of heaven	Coconut

❑❑❑

Appendix-4

TERM AND ASSOCIATED CROPS

S.No	Terms	Crop	Remarks
1.	Mattocking	Banana	Sequential cutting of the plant after harvest
2.	De-suckering	Banana	Desuckring once in 45 days
3.	Denavelling	Banana	Removal of male bud after completion of the female phase
4.	Propping	Banana	Essential for tall cultivars
5.	Metaxenia	Date palm and Papaya	Pollens have a direct influence of somatic tissue of fruit oseedutside the embryo and endosperm results in early or late ripening, change in size and colour of fruits and seed
6.	Doka stage	Date palm	Harvesting stage of date palm before onset of rains
7.	Caprification	Fig	Pollination of common fig with caprifig
8.	Notching	Fig	It is involves removing of small slice of bark immediately above the dormant bud, giving two slanting cuts as deep as the bark
9.	Ikebana	Flowers	Technique of flower arrangement
10.	Terrariums	Gardening	Growing plants in terraces
11.	Girdling	Grapes	It increases the berry size by 1 to 1.5 mm
12.	Root pruning	Guava	Crop regulation
13.	Medo orcharding	Guava	Planting of plants at 2 1 meter distance
14.	Bahar treatment	Guava, Pomegranate, Citrus fruit	To regulate flowering and fruiting
15.	Smudging	Mango	To induction flowers in mango
16.	Caging technique	Mango	Mango breeding
17.	Viviparous	Jackfruit	Germination seed on mother plant

MEMORY BASED QUESTIONS

A. Ornamental and Landscaping & MAP

1. MAP is study of
   (A) fruit (B) Mushroom
   (C) flower (D) Medicinal plants
2. **Withaniasominifera is botanical name of**
   (A) Aswagandha (B) Satavar
   (C) Safedmusli (D) Sarpagandha
3. **Rauwolfiaserpentina is harvested after**
   (A) 5-10 months (B) 16-20 months
   (C) 20-24 months (D) 12-15 months
4. **Menthe (mints) is**
   (A) Long day plant (B) Short day plant
   (C) Day neutral pant (D) None
5. **Inflorescence in mints is**
   (A) Umbel (B) Spikes
   (C) Raceme (D) Cymose
6. **Khus-Khus is a native of**
   (A) India (B) Central America
   (C) Mexico (D) Asia
7. **Davana is a native of**
   (A) China (B) India
   (C) Mexico (D) Japan
8. **Botanical name of celery**
   (A) Abelomoschus moschatus (B) Apiumgraveolens
   (C) Pelargonium graveolens (D) None

9. **Family of lemon grass**
   (A) Solanaceae (B) Chenopodiaceae
   (C) Malvaceae (D) Poaceae
10. **Patchouli is propagated through**
   (A) Seed (B) Division
   (C) Layering (D) Terminal shoot cutting
11. **Patchouli (Pogostem on patchouli) is**
   (A) Short Day plant (B) LDP
   (C) DN (D) None
12. **Which country the largest producer of patchouli oil**
   (A) India (B) Indonesia
   (C) China (D) Japan
13. **Native place of Ocimum**
   (A) India (B) Mexico
   (C) South Africa (D) none
14. **Eucalyptus belongs to family**
   (A) Lamaceae (B) Rosaceae
   (C) Santalaceae (D) Myrtacae
15. **Santalum album is a botanical name of**
   (A) Lavander (B) eucalyptus
   (C) Sandalwood (D) Lamiaceae
16. **Which biomolecules Is responsible for anticancer activity**
   (A) Glucose (B) Fructose
   (C) Flavanoids (D) Calcuim
17. **Which of the following is not the property of Phyllanthusniruri?**
   (A) Anti cancer (B) Cataract
   (C) Jaundice (D) Anti oxident
18. **Which of the practitioners make use of medicinal plant ?**
   (A) Yoga (B) Acupressure
   (C) Ayurveda (D) Acupuncture
19. **What is the name given to the chemicals extracted from plant?**
   (A) Biochemicals (B) Phytochemicals
   (C) Molecular chemicals (D) none

20. Which of following extraction technology involves continuous mode extraction ?

(A) Cold percolation
(B) Maceration
(C) Agitation
(D) Microwave Assisted Extraction

21. Which of the following extraction technology is solid phase extraction technology?

(A) Cold percolation
(B) Maceration
(C) Agitation
(D) None of these

22. Which extraction technology has smaller solvent to solid ratio for taxol extraction from Taxusbrevifolia

(A) Cold percolation
(B) Maceration
(C) Agitation
(D) Microwave Assisted extraction

23. Which of the following technology use carbon carbon dioxide as a solvent?

(A) Cold percolation
(B) Maceration
(C) SCFE
(D) Microwave Assisted extraction

24. Which of the following is used as an enhancer for SCFE?

(A) Microwave
(B) Ultrasound
(C) Both A and B
(D) None

25. Why drying temperature and time should be less?

(A) To prevent degradation of the bioactive compounds
(B) To prevent evaporation of the bioactive compounds
(C) To prevent condensation of the bioactive compounds

26. Fragrance is controlled by

(A) Polygene
(B) Oligogene
(C) Monogene
(D) None of these

27. Money plant is also known as

(A) Ficusrepens
(B) Hederahalix
(C) Irish ivy
(D) Scindapsisaureus

28. Which is the economic part of Guggal

(A) Leaves and fruits
(B) Roots
(C) Gum resin
(D) None of these

29. Economic part of sarpagandha is

(A) Dried root
(B) Seed
(C) Leaves
(D) None of these

30. Opium is used as

(A) Laxative (B) Asthma

(C) Painkiller (D) Intestinal disorder

31. Kalmegh is used in

(A) Snake bites (B) Dog bites

(C) Antistress (D) Controlling blood pressure

32. Sarpagandha is used in

(A) Painkiller (B) Anti-diabetic

(C) Jaundice (D) Contraceptive pills

33. Java citronella is used in

(A) Painkiller (B) Appetizer

(C) Laxative (D) Mosquito repellent and deodorants

34. National board of Aromatic and medicinal is situated at

(A) Mumbai (B) Nagpur

(C) Lucknow (D) New Delhi

35. In the following most common laxative is

(A) Opium (B) Lobelia

(C) Plantago (D) Periwinkle

36. CIMAP is located at

(A) New Delhi (B) Nagpur

(C) Hyderabad (D) Lucknow

37. In the following which coniferous tree is known to cure cancer

(A) Ferula narthex (B) Cassia fistula

(C) Taxusbaccata (D) Terminaliabellerica

38. Plant grown in the alpine foresrs is

(A) Aloevera (B) Cassia fistula

(C) Podophyllumhexandrum (D) Piper longum

39. In the following which is commonly known as sweet herb for the use of diabetic herb

(A) Cassia (B) Stevia

(C) Centella (D) Neem

40. Which plant is yielding drug from all its plant parts

(A) Poppy (B) Aloe

(C) Ocimum (D) Psyllium

41. In the following medicinal plants ,which has maximum acreage in india

(A) Digitalis (B) Mint

(C) Safedmusli (D) neem

42. Which plant is commonly used to cure cardiovascular disease

(A) Digitalis (B) Mint

(C) Safedmusli (D) Neem

43. Which is a branch of medical science dealing with the action of drug

(A) Nematology (B) Pharmacology

(C) Pharmacognosy (D) Aromatherapy

44. For the treatment of malaria,quinine, is extracted from

(A) Cinchona (B) Poppy

(C) Aonla (D) Celery

45. Violarsurpens is useful in curing

(A) Asthma (B) Vomiting

(C) Cough (D) Headache

46. Spilanthes is useful in

(A) Toothache (B) Cough

(C) Dysentery (D) None of these

47. Plant commonly used to cure cardiovascular disease is

(A) Aloe (B) Mint

(C) Ferula (D) Digitalis

48. Teapentine drug is obtained from

(A) Pine (B) Celery

(C) Poppy (D) Aloe

49. Which of the following plant is not yielding drug from its fruits

(A) Aonla (B) Bael

(C) Garlic (D) None of these

50. Bael belongs to family

(A) Rutaceae (B) Liliaceae

(C) Apocynaceae (D) Asteraceae

51. Botanical name of aloe

(A) Aloevera (B) Mangiferaindica

(C) Psidiumguajava (D) Citrus limonica

52. Family of aloe

(A) Liliaceae (B) Umbelifereae

(C) Rutaceae (D) Crucifereae

53. Propagation is done through aloe

(A) Root suckers / Rhizome cutting (B) Runner

(C) Air layering (D) Gooty

54. Which varieties of aloe

(A) IC 111271 (B) NA 5

(C) NA 10 (D) NA 7

55. Major producers state of aloe vera in india

(A) Rajasthan (B) tamilnadu

(C) Mumbai (D) Uttarpradesh

56. Botanical name of babchi is

(A) Psoraleacorylifolia (B) Crotolareajunsia

(C) Symblicaofficinal (D) Solanumtuberosum

57. Chromosome number of the babchi is

(A) 2n = 20 (B) 2n = 18

(C) 2n = 27 (D) 2n = 90

58. Orogin place of the babchi

(A) India (B) china

(C) Pakistan (D) jammuamdkashmir

59. Propagated through the babchi

(A) Seeds (B) Layering

(C) Cutting (D) Transplant

60. Which alkaloids available for the babchi.

(A) Sanigrin (B) Coumarins

(C) Mmarmelosin (D) Lycopin

61. Botanical name of belladonna is

(A) Atropabelladona (B) Atropaitalika

(C) Atropasurpentina (D) Zingiberpfficinale

62. Family of the belladonna is

(A) Solanaceae (B) Umbilifereae

(C) Cucurbitaceae (D) Rutaceae

63. Origin place of the belladonna is

(A) Europe and Asia
(B) America
(C) Japan
(D) China

64. Chromosome number of the belladonna is

(A) 2n = 33
(B) 2n = 40
(C) 2n = 38
(D) 2n = 72

65. Belladonna is treatment for

(A) Gout disease
(B) Rheumatism
(C) Parkinson's
(D) All

66. Belladonna is propagated by

(A) Layering
(B) Cutting
(C) Budding
(D) Seed

67. Belladonna leaves are widely used for manufacture of

(A) Plasters
(B) Tinctures
(C) Extracts
(D) All

68. Chemical compounds found in Belladonna

(A) Hyosicne
(B) Atropine
(C) Both
(D) None

69. Botanical name of Dill

(A) Papaversomniferum
(B) Glycrrhizaglabra
(C) Anethumgraveolens
(D) None

70. Family of Dill

(A) Scrophularaceae
(B) Apiaceae
(C) Burceraceae
(D) none

71. Common pest of Dill

(A) Powdery mildew
(B) Damping off
(C) CMD
(D) ALL

72. Dill is also known as

(A) Henbene
(B) Solva
(C) Both A & B
(D) None

73. Botanical name of Fox Glove

(A) Digitalis lantan
(B) Glycrrhizaglobra
(C) Anethum Spp.
(D) None

74. Family of Fox Glove
(A) Apiaceae (B) Scrophularaceae
(C) Burseraceae (D) All

75. Fox Glove propagated through
(A) Seed (B) Tissue culture
(C) Budding (D) Layering

76. Guggal is also known as
(A) Dill (B) Leaves
(C) Indian Bdellium tree (D) None

77. Botanical name of Guggal
(A) Commiphorawightii (B) Hyocyamusniger
(C) Atropa belladonna (D) None

78. Family of Guggal
(A) Rehmannaceae (B) Lauraceae
(C) Burseraceae (D) Papaveraceae

79. Guggal are
(A) Shrub (B) Small tree
(C) Herb (D) Both A&B

80. Economic part of Guggal
(A) Gum resin (B) Seed
(C) Leaves (D) None

81. Guggal is used fortreatment
(A) Arthritis (B) Obesity
(C) Both A&B (D) None

82. Guggal propagated through
(A) Hard wood cutting (B) Semi wood stem cutting
(C) layering (D) Budding

83. Tapping is done for
(A) Mulethi (B) Papaver
(C) Guggal (D) All

84. Botanical name of Henbene
(A) Hyocyamusniger (B) Glycrrhizaglabra
(C) Withaniasomnifera (D) None

85. CIMAP is working under which authority

(A) CSIR (B) ICAR

(C) ASRB (D) NBAMP

86. Seed rate per hectare of Ashawgandha is about

(A) 8-10 Kg (B) 5-6 Kg

(C) 3-4 Kg (D) 2 Kg

87. Sarpgandha is commercially propagated by

(A) Root cutting (B) Layering

(C) Seeds (D) None

88. Opium is sowing at which month

(A) Oct-Nov (B) Nov-Dec

(C) Dec-Jan (D) Jan- Feb

89. Mentha can be propagated by

(A) Runner (B) Sucker

(C) Seed (D) Cutting

90. Aloe Vera is origin from

(A) India (B) Mexico

(C) Africa (D) None

91. Lemon grass is propagated by

(A) Slips (B) Runner

(C) Sucker (D) Cutting

92. Seed rate of Isabgol is

(A) 5-6 Kg/ha. (B) 1-3Kg/ha.

(C) 2 Kg/ha. (D) None

93. Husk and Seed ratio in Isabgol found is about

(A) 30:20 (B) 30:40

(C) 40:50 (D) 30:70

94. Isabgoloriginatelly from

(A) India (B) Africa

(C) China (D) None

95. Withaniasomnifera is the botanical name of

(A) Satawar (B) SafedMusli

(C) Sarpgandha (D) None

96. Crop duration of Rauwelfia serpentine is about

(A) 9-10 months (B) 14-16 months

(C) 18-20 months (D) None

97. Mentha (Mints) is

(A) SDP (B) LDP

(C) DNP (D) None

98. Khuskhus native to

(A) India (B) Mexico

(C) North America (D) Asia

99. Lemon grass belongs to

(A) Poaceae (B) Gramineae

(C) Both (D) Solanaceae

100. Economical part of Sarpgandha is

(A) Dried leaves (B) Dried roots

(C) Mature stem (D) Immature stem

101. Leading country with production of ocimum oil

(A) India (B) Japan

(C) China (D) Indonesia

102. National Board of Aromatic and Medicinal Plant is situated at

(A) Bihar (B) Andhra Pradesh

(C) Tamil Naidu (D) None

103. CIMAP is situated at

(A) Mumbai (B) Ranchi

(C) Hyderabad (D) None

104. Sowing time of Opium poppy

(A) Oct-Nov (B) July-August

(C) Feb-March (D) May-june

105. Seed rate of opium poppy

(A) 3 kg (B) 5 kg

(C) 8 kg (D) 10 kg

106. Arkaupkar is the variety of

(A) Sarpagandha (B) dioscorea

(C) Opium poppy (D) Alovera

107. Sennabelong to family

(A) Caesalpinaceae (B) Solanaceae

(C) Liliaceae (D) Acanthaceae

108. Senna is mainly propagated by

(A) Budding (B) Cutting

(C) Seed (D) Non of these

109. Chromosome number (2n) of senna

(A) 20 (B) 16

(C) 24 (D) 34

110. Senna plant well grown in

(A) Loamy soil (B) Sandy loam soil

(C) Alkaline soil (D) Non of above

111. Senna is originated by from

(A) Saudi Arbia& India (B) Burma

(C) America (D) Japan

112. Aswgandha belong to family

(A) Compositae (B) Asteraceae

(C) Leguminosae (D) Solanaceae

113. Aswgandha plant thrive well in soil

(A) Clay (B) Sandy

(C) Clay loam (D) Water logged soil

114. The Botanical Name of Aswgandha is

(A) Withaniacompecta (B) Withania sativa

(C) Papoversomnifera (D) Withaniasomnifera

115. The economical part of Aswgandha is

(A) Stem (B) Leaves

(C) Dried roots (D) None of these

116. Aswgandha crop planted in month of

(A) March – April (B) October - November

(C) July (D) August – September

117. The pathogen associated with chlorophytom anthracnose is

(A) Choletotrichum spp. (B) Phytopthorea spp.

(C) Fuarium spp. (D) August – September

118. Chlorosis of the safedmusli is due to deficiency of

(A) Ferrous (B) Sulphate

(C) Chloride (D) None of these

119. Safedmusli is commonly used for

(A) Panicle (B) Dried roots

(C) Leaves (D) Pods

120. The practice is done to increase the tuber

(A) Topping (B) Pinching

(C) Desuckering (D) None of the above

121. The chemical compound found in Sadabahar

(A) Giraniol (B) Pinching

(C) Desuckering (D) Cristisene

122. The chemical compound found in Sarpagandha

(A) Giraniol (B) Pinching

(C) Reserpine (D) Cristisene

123. The chemical compound found in Chandan

(A) Giraniol (B) Renol

(C) Santolol (D) Cristisene

124. The chemical compound found in Mulethi

(A) Glycyrrhizine (B) Renol

(C) Santolol (D) Resin

125. The chemical compound found in Rose-

(A) Glycyrrhizine (B) Giraniol

(C) Santolol (D) Resin

126. The chemical compound found in Jawacitronila

(A) Glycyrrhizine (B) Giraniol

(C) Santolol (D) Citronellol

127. The chemical compound found in khus

(A) Khushinol (B) Giraniol

(C) Santolol (D) Citronellol

128. The chemical compound found in Lemongrass

(A) Khushinol (B) Citrol

(C) Santolol (D) Citronellol

129. The chemical compound found in Mentha

(A) Khushinol
(B) Methyl acetate
(C) Menthol
(D) Both B & C

130. The chemical compound found in Pipli

(A) Resin
(B) Volatile oil
(C) Piperine
(D) All

131. The chemical compound found in Ashwagandha

(A) Resin
(B) Withanine
(C) Piperine
(D) All

132. The chemical compound found in Kalmegh

(A) Resin
(B) Withanine
(C) Piperine
(D) Andrographiolite

133. The chemical compound found in Satawar

(A) Resin
(B) Withanine
(C) Piperine
(D) Saponin

134. The chemical compound found in Isabgol

(A) Resin
(B) Withanine
(C) Piperine
(D) None of these

135. The chemical compound found in senna

(A) Resin
(B) Sanosides
(C) Piperine
(D) Saponin

136. The chemical compound found in Kalihari

(A) Cheublic acid
(B) Colchicine
(C) Piperine
(D) None of these

137. The chemical compound found in Baheda

(A) Resin
(B) Chebulic acid
(C) Chebulin
(D) None of these

138. The chemical compound found in Harad

(A) Chubilin
(B) Behenic acid
(C) Piperine
(D) Both A & B

139. The chemical compound found in Kauch

(A) Resin
(B) Withanine
(C) Piperine
(D) None of these

140. The chemical compound found in Kauch

(A) Mucilage (B) Polysaccharides

(C) L-dopa (D) Both A & B

141. The chemical compound found in Brahmi

(A) Mucilage (B) Polysaccharides

(C) Both A&B (D) None of these

142. The chemical compound found in Pattharchur

(A) Mucilage (B) Polysaccharides

(C) Carvocrol (D) Colchicine

143. The chemical compound found in Brahmi

(A) Mucilage (B) Polysaccharides

(C) Both A&B (D) Brahminoside

144. The chemical compound found in Pipli

(A) Resin (B) Polysaccharides

(C) Piperine (D) Both A & C

145. The chemical compound is not found in mentha

(A) Menthol (B) Menthone

(C) Methyle Acetate (D) Citrol

146. The chemical compound is not found in Opium poppy

(A) Morphine (B) Codine

(C) thebane (D) Reserpine

147. The chemical compound is not found in chandan

(A) Volatile oil (B) Resin

(C) Santolol (D) Carvocrol

148. The chemical compound is not found in Pipli

(A) Resin (B) Polysaccharides

(C) Piperine (D) Volatile oil

B. Growth and Development of Horticultural Crops.

1. Which of the following plant hormones is used for berry elongation in graps

(a) IAA (b) ABA (c) GA_3 (d) NAA

2. Plant growth regulator used for fruit thinning

(a) IAA (b) NAA (c) GA_3 (d) C_2H_4

3. Which growth regulator is used to reduce period of chilling requirement

(a) Cytokinin (b) IBA (c) NAA (d) GA_3

4. Plant growth regulator used for rooting in cutting

(a) ABA (b) NAA (c) IBA (d) GA_3

5. Fruit which yellow in colour are rich source of

(a) Vitamin A (b) Vitamin B

(c) Vitamin C (d) Vitamin E

6. Mangosteen fruit contains seed and development of process of seed formation is called

(a) Parthenocarpy (b) Polyembryony

(c) Parthenogenesis (d) Apomixis

7. A series of biochemical changes initiated by the catalytic production of ethylene making the changes from growth to senescence and involving a increase in respiration rate is called

(a) Maturity (b) Senescence (c) Climatric (d) Ripening

8. Which harmones is more effecting in ripening fruit crop

(a) GA_3 (b) Kintenine (c) IAA (d) Ethephon

9. AVG is a inhibitor is

(a) Cytokine (b) Auxine (c) Ethylene (d) GA_3

10. On tree storage in citrus is possible with application of

(a) 2-4-D (b) Thiourea (c) ABA (d) GA_3

11. In grape which training system gives maximum cost benefit ratio

(a) Kniffine (b) Bower (c) Teliphon (d) Head

13. Which of the following is used to check sprouting for onion during storage

(a) NAA (b) MH (c) GA_3 (d) PCPA

14. Flower initiation and higher yield of onion bulbs may be expected with the treatment of

(a) NAA (b) IAA (c) IBA (d) GA_3

15. Sprouting of potato can be controlled by a pre harvest spray of

(a) IBA (b) MH

(c) CCC (d) Non of the these

16. To induce flowering growth in sweet potato regulator is used

(a) IBA (b) ABA (c) GA3 (d) 2-4-D

17. Growth regulator used for fruit setting in short style of flowers of brinjal is

(a) ABA (b) IAA (c) NAA (d) 2-4-D

18. Which of the following is generally used to break the dormancy in potato

(a) GA_3 (b) NAA (c) Ethylene (d) ABA

18. Which of the following plant harmones is also known as Phytokinin

(a) Cytokinine (b) GA3 (c) Auxine (d) ABA

19. Naturally occurring cytokinin is

(a) Kinetin (b) Zeatine
(c) Adanin (d) None of the these

20. Dormancy of corm in gladiolus is due to

(a) 2-4-D (b) NAA (c) ABA (d) IBA

C. Plant Propagation and Nursery Management

1. Fill in the blanks:-

1. Commercial method of propagation in apple............................
2. Commercial method of propagation in apricot ....
3. Commercial method of propagation in almond ....
4. Commercial method of propagation in peach ....
5. Commercial method of propagation in pear ....
6. Commercial method of propagation in plum ....
7. Commercial method of propagation in pecannut ....
8. Commercial method of propagation in walnut ....
9. Commercial method of propagation in quince ....
10. Commercial method of propagation in hazalnut ....
11. Commercial method of propagation in Black raspberry ....
12. Commercial method of propagation in cape gooseberry ....
13. Commercial method of propagation in cherry ....
14. Commercial method of propagation inchestnut ....
15. Commercial method of propagation in Strawberry ....
16. Commercial method of propagation in Surenam cherry .... grafting.
17. Commercial method of propagation in Barbados cherry................ layering.
18. Commercial method of propagation in pista ....
19. Commercial method of propagation in Chilogoza ....

20. Commercial method of propagation in grape ....
21. Commercial method of propagation in cashewnut ....
22. Ultra dwarf root stock of apple ....
23. propagation in apple during .... and ....
24. Whip grafting I young plants having diameter of .... cm.
25. Inarching is also known as .... grafting.
26. Veneer grafting is occurring in ....
27. Ring budding is occurring in ....
28. .... is important components of hi-tech horticulture for propagation of high value horticultural crops.
29. Air layering is also known as ....
30. In .... rapidly for quick dissemination of disease free planting material
31. Rooting hormone ....
32. Use of leaf and leaf parts for regeneration of new plants called .... cutting.
33. Use of shoot cutting with older wood called .... cutting.
34. Use of shoot cutting with small piece of older wood called .... cutting.
35. Cutting without older wood containing bud eyes.................... cutting
36. Give the name of any layering ....
37. .... is the processby which the plants are conditioned with stand against adverse conditions after transplanting in the field.
38. Dwarf root stock of pear ....
39. Dwarf root stock of peach ....
40. Dwarf root stock of cherry............................
41. .... is the parts of graft which provides root system to the grafted plant.
42. .... Is the part of graft which is used in between rootstock and scion variety for successful graft.
43. The commercial rootstock are divided as .... and .... stock ....
44. Rootstock of apricot ....
45. Rootstock of guava ....

46. Root stock of ber ....
47. Commercial method of propagation in litchi ....
48. Commercial method of propagation in sapota ....
49. Commercial method of propagation in persimon ....
50. Commercial method of propagation in avocado ....
51. Commercial method of propagation in wood apple ....
52. Commercial method of propagation in mangosteen ....
53. Commercial method of propagation in papaya ....
54. Commercial method of propagation in mandarin ....
55. Commercial method of propagation in loquat ....
56. Commercial method of propagation in karonda ....
57. Commercial method of propagation in phalsa ....
58. Commercial method of propagation in tamarind ....
59. Commercial method of propagation in pomegranate ....
60. Commercial method of propagation in pineapple ....
61. Commercial method of propagation in banana ....
62. Commercial method of propagation in aonla ....
63. Commercial method of propagation in jackfruit ....
64. Time of propagation of peacannut ....
65. Time of propagation of walnut ....
66. Time of propagation of apple ....
67. Time of propagation of apricot ....
68. Time of propagation of almond ....
69. Time of propagation of peach ....
70. Time of propagation of pear ....
71. Time of propagation of plum ....
72. Time of propagation quince ....
73. Time of propagation of hazzalnut ....
74. Time of propagation of black raspberry ....
75. Time of propagation of red raspberry ....
76. Time of propagation of cape gooseberry ....
77. Time of propagation of chestnut November ....

78. Time of propagation of strawberry ....
79. Time of propagation of surenam cherry ....
80. Time of propagation of barbadose cherry ....
81. Time of propagation of pista ....
82. Time of propagation of chilgoza ....
83. Time of propagation of grape ....
84. Time of propagation of craneberry............................
85. Time of propagation cashewnut ....
86. Time of propagation of coconut ....
87. Time of propagation of date palm ....
88. Time of propagation of mango ....
89. Time of propagation of guava ....
90. Time of propagation of ber ....
91. Time of propagation of Bael ....
92. Time of propagation of jackfruit ....
93. Time of propagation of aonla ....
94. Time of propagation of banana ....
95. Time of propagation of pineapple ....
96. Time of propagation of custard apple ....
97. Time of propagation of sapota ....
98. Time of propagation of pomegranate ....
99. Time of propagation of tamarind ....
100. Time of propagation of litchi ....
101. Time of propagation of phalsa ....
102. Time of peropagation of karonda ....
103. Time of propagation of loquat ....
104. Time of propagation of sweet orange ....
105. Time of propagation of mandarine ....
106. Time of propagation of acid lime ....
107. Time of propagation of lemon ....
108. Time of propagation of papaya ....
109. Time of propagation of Mangosteen ....

110. Time of propagation of wood apple ....
111. Time of propagation of avocado ....
112. Time of propagation of persimon ....
113. Time of propagation of paniala ....
114. Night blindness is caused due to deficiency of vitamin ....
115. Venturiainaequalis is the causal organism of ....
116. Black tip of mango occur due to toxicity of ....
117. Saracaindica is the botanical name of ....
118. Jasminum belongs to family ....
119. Mango cultivar-alphonso is more susceptible to ....
120. Edible part of turmeric is ....
121. Cabbage is classified as .... season crop.
122. Bhaduran-variety of mango is tolerant to ....
123. Walnut is the richest source of ....
124. The best pruning time for rose inn.india is .... month.
125. Gladiolus is propagated by ....
126. Radish is a .... plant.
127. Browning in cauliflower occurs due to .... deficiency.
128. Application of fertilizers with irrigation water is known as ....
129. Strawberry grows well in .... climate
130. .... is the abbreviated form of food product act.
131. .... is used for pectin test in fruit extracts.
132. Papain is prepared from .... fruits
133. Chilli fruit is botanically ....
134. Removal of growing point of a shoot along with few leaves is known as ....
135. Clerodendroninerme-is the most common shrub for evergreen ....
136. Fruit preservation and canning institute at lucknow was established in the year - ....
137. Pusabedana is a seedless hybrid of ....
138. Pusabedana is a cross of tetra 2 ....
139. Althearosea is a winter season ....
140. The annual yield of tomato crop is ....quintal per hectare.

141. Beal belong to the family............................
142. ....is the dwarfing root stock of mango .
143. Ber should be pruned during .... month
144. .... is an off season cultivar of mango.
145. Bitter gourd is a .... plant by sex form
146. Okra has back cross somatic chromosomes ....
147. Banana is commercially propagated by ....
148. Okra has 2n = ... sometic chromosomes.
149. Banana is commercially propagated by ....
150. .... is commercial method of aonla propagation.
151. .... state is the largest producer of guava.
152. A cross between a hybrid and one of its parent is known as ....
153. .... is the origin centre of cowpea.
154. The superiority of f1 hybrid over both of its parents is known as ....
155. Botanically ber fruit is a ....
156. Litchi is commercially propagated by ....
157. Cassia fistula is the botanically name of ....
158. China aster is commercially propagated by ....
159. Ficusglomerata is nematode resistant root stock of ....
160. Phalsa is the native of ....
161. Pusaharit is the variety of ....
162. An uninjured garlic clove contains .... amino acid.
163. Hardwood cutting is the commercial method of .... propagation.
164. Eeiobotryajaponica is the native of ....
165. ....is the largest producer of pear in the world.
166. Generally fruit cropps prefer a ph range of ....
167. ....is the largest producer of walnut in India.
168. Artocarpusheterophyllus is the native of ....
169. Mango has 2n=40 .... chromosome number.
170. Seed rate of cauliflower is .... gram per hectare.
171. Pant c-1 chilly is tolerant to ....
172. Dwarf banana should be planted at ....mete.

173. ……….. cultivar of mango is suitable for high density planting ……………..
174. J.grandiflorum species of jasminum is suitable for ……………………….
175. ……………… method is the most common system of planting.
176. …………………… is major insect of mango.
177. …………………… and …………………… are poly embryonic variety of mango.
178. Use of …………………… is effective method to control mango mealy bug.
179. Ripe fruit of …………………… variety of mango remain green in colour.
180. Fruit cracking is serious problem in …………………… .
181. Name of Grapefruit is because of …………………… .
182. Citrus canker is serious problem of …………………… .
183. Lemon is propagated commercially by…………………… .
184. Thompson seedless is …………………… pruned variety of grape.
185. …………………… is most suitable variety for raisin making.
186. Chromosome number of Vitisvinifera is …………………… .
187. Pruning of spur type grape variety should be done leaving …………………… buds.
188. Annonafruit belongs to family…………………… .
189. …………………… is the type of fruit of Custard apple.
190. Family of Mangosteen is …………………… .
191. NB-9 is a variety of …………………… .
192. Botanical name of Rambutan is …………………… .
193. Chromosome no. of Bael is …………………… .
194. Family of Durian is …………………… .
195. Aonla is commercially propagated by…………………… .
196. Edible part of Bael is …………………… .
197. Durian belong to family …………………… .
198. Family of Mangosteen is …………………… .
199. Family of Avocado is …………………… .
200. Write any two varieties of Rambutan …………………… and …………………… .
201. Rambutan is commercially propagated by …………………… .
202. Origin place of Rambutan is …………………… .
203. Bael is commercially propagated by …………………… .

204. Edible part of Aonla is ....
205. Type of inflorescence in Mango is ....
206. Botanical name of Aonla is ....
207. Edible part of Annona is ....
208. Predominant acid present in Peach is ....
209. Flower color of Pear is ....
210. .... fruit is plant symbol of Lord Shiva.
211. Type of fruit in Fig is ....
212. Predominant pollination in Pomegranate done by ....
213. Strawberry Chromosome number (2n) = ...
214. .... is responsible for flavor of Strawberry.
215. Type of inflorescence in Aonla is ....
216. Botanical name of Rose apple is ....
217. Rose apple belongs to family ....
218. Chromosome no. of Avocado is ....
219. Mango is a .... type of fruit.
220. Type of inflorescence in Mango is ....
221. .... is major disease of Citrus.
222. .... and .... are variety of Guava.
223. Use of .... is effective method to control Citrus canker.
224. .... cultivar of Citrusis used for juice making at maturity stage.
225. Tristeza is serious problem in ....
226. Leaves of .... fruit is used as offerings to lord Shiva.
227. Malformation is serious problem of ....
228. Mandarin is propagated commercially by........................ .
229. Anab-e-shahi is .... pruned variety of grape.
230. .... cultivarof banana is most suitable for table purpose.
231. Chromosome number of Mangiferaindica is ....
232. Pruning of cane type grape variety should be done leaving .... buds.
233. Aonlafruit belongs to family ....
234. .... is the type of fruit of Mango.
235. Family of Rose apple is ....

236. Beauty seedless is a variety of ....
237. Botanical name of Carambola is ....
238. Chromosome no. of Aonla is ....
239. Botanical name of Durian is ....
240. Mango is commercially propagated by ....
241. Edible part of Citrus is ....
242. Sapotabelong to family ....
243. Family of Litchi is ....
244. Family of Passion fruit is ....
245. Write any two varieties of Grape .... and ....
246. Litchi is commercially propagated by ....
247. Origin place of Grape is ....
248. Bael is commercially propagated by ....
249. Edible part of Litchi is ....
250. Type of fruit in Grape is ....
251. Botanical name of Bilimbi is ....
252. Edible part of Papaya is ....
253. Predominant acid present in Citrus is ....
254. Flower color of Bael is ....
255. .... fruit is known as queen of fruits.
256. Type of fruit in Guava is ....
257. Pineappleis Commercial propagated by ....
258. Sapota Chromosome number (2n) = ...
259. Black tip is a physiological disorder of ....
260. Type of fruit in pomegranate is ....
261. Botanical name of Sapota is ....
262. Breadfruit belongs to family ....
263. Chromosome no. of Bilimbi is ....
264. Loquat is a .... type of fruit.
265. Type of fruit in Durian is ....
266. .... fruit is commonly known as king of fruits.
267. Chakaiyaand NA-10 are varieties of ....

268. Use of bordeux mixture is effective method to control ....
269. Mosambicultivar of Citrusis used for juice making at .... stage.
270. Tristeza is serious problem in ....
271. .... of Bael fruit is used as offerings to lord Shiva
272. Malformation is .... of Mango.
273. Mandarin is propagated commercially by ....
274. Beauty seedless is .... pruned variety of grape.
275. .... cultivarof banana is most suitable for vegetable purpose.
276. Chromosome number of Mangiferaindica is ....
277. Family of Litchi is ....
278. Family of .... is Passifloraceae
279. Write any two varieties of Grape .... and ....
280. Lemon is propagated commercially by ....
281. .... is spur pruned variety of grape.
282. .... is most suitable variety for raisin making.
283. Chromosome number of Vitisvinifera is ....
284. Pruning of spur type grape variety should be done leaving .... buds.
285. .... belongs to familyAnacardiaceae.
286. .... is the type of fruit of Citrus.
287. Family of Mangosteen is ....
288. .... is a variety of Bael.
289. Litchi is commercially propagated by ....
290. Origin place of Grape is ....
291. Bael is commercially propagated by ....
292. Edible part of Litchi is ....
293. Bananais Commercial propagated by........................ .
294. Type of fruit in .... is drupe.
295. Botanical name of Bilimbi is ....
296. Edible part of Papaya is ....
297. Predominant acid present in .... is tartaric acid.
298. Flower color of Bael is ....
299. .... fruit is known as queen of fruits.

300. Type of fruit in Guava is Balausta.
301. Pineappleis Commercial propagated by ....
302. The best planting material in strawberry is ....
303. Phalsa is comerciallypropageted by ....
304. Pineapple is commercially propagated by ....
305. Gladiolus propagated by ....
306. Commeriallypropagetion of Bougainvillea ....
307. papaya is commercially propagated by ....
308. Hard wood cutting can be used to propagated fruit crop ....
309. Grape is commercially propageted by ....
310. Guava is commercial propageted by ....
311. Phalsa is commercial propageted by ....
312. Cerries is commercially propageted through ....
313. National flower of india is ....
314. Queen of flower is ....
315. Dahlia is propagated by ....
316. Ber is commercially propagated by ....
317. Karonda is commercially propagated by ....
318. Dwarfing rootstock of apple is ....
319. Standardized rootstock of guava is ....
320. Semi hard wood cutting can be used to .... fruitcrope.
321. Litchi is commercially propagated by ....
322. Vellaikolamban is a root stock of ....
323. Most commercial method of propagation of mango in western U.P. ....
324. Coconut is commercially propagated by ....
325. Date palm is commercially propagated by ....
326. Grape is commercially propagated by .... cutting
327. Grape fruit is commercially propagated by .... budding
328. In situ method of grafting in mango...........................
329. Which type of budding is commercially used in almond ....
330. Which type of grafting is commercially used in cahew nut ....
331. Papaya is commercially propagated by ....

332. What is the botanical name of chinese guava used as rootstock ....

333. Name the dwarfing root stock of guava ....

334. Which type of grafting is used in persimmon fruit ....

335. Strawberry is commercially propagated by..................................

336. Which is method of tissue culture to get plant free from virus..................

337. Formation of haploid plants can be done by ....

338. Lower portion of grafted planted is called ....

339. Which is extreme dwarfing rootstock of apple ....

340. Sweet orange is commercially propagated by..................................

341. Pineapple is commercially propagated by ....

342. ....is most promising rootstock of mandarin and sweet orange in central and south India

343. Promising dwarf rootstock of guava which is also known as aneuploid 82 ....

344. Quince rootstock are belongs to ....

345. Quince C is dwarfing rootstock of ....

346. Semi vigorous rootstock of pear is ....

347. GF – 305 is a rootstock of ....

348. Siberian C is .... resistant rootstock of peach

349. Nemared is .... resistant rootstock of peach

350. Colt is common root stock of ....

351. Semi dwarfing rootstock of cherry..................................

352. Colt is a rootstock of cherry cross of .... And ....

353. Standard rootstock of cherry for both sweet and sour cherry ....

354. Paradox is a vigorous rootstock of ....

355. Myrobalan GF 31 is a rootstock of ....

356. Caryaacquiaticais commonly used rootstock of ....

357. Propagating a new plant from a piece or part of a mother plant is called as .... propagation

358. Marcottage is another name of ....

359. Trifoliate orange is resistant to .... and ....

360. Kinnow variety was introduced in India in which year ....

361. Kinnow can grown in high density planting by using .... as a root stock

362. Root stock adajamir is resistant to ....
363. Best rootstock for mosambi ....
364. .... is best rootstock for Rough lemon
365. Ultra dwarf root stock of citrus
366. Inarching is also known as .... grafting
367. East malling is rootstock of ....
368. Technique of produced genetically identical individuals from a single cell is known as ....
369. Which is commercial method of propagation of loquat ....
370. Tongue grafting is the modified form of ....
371. .... means growing particular plants parts in darkness.
372. What is the temperature of bottom heat chamber ....
373. Sectorial chimera is .... type
374. Most stable form of chimera ....
375. A plant or plant parts composed of genetically different layers is known as ....
376. Root cuttings are practised in fruit crops ....
377. Commonly used medium in air layering ....
378. Ideal period (month) for air layering ....
379. Grafting which is most ideal for establishing insitu orchards and top working of old orchards ....
380. Scarification means ....
381. Temperature of cryopreservation of seeds ....
382. Name the commonly use cryoprotectants ....
383. Polyembryony term was coined by ....
384. Polyembryony first discovered by ....
385. Cleavage embryony is found in ....
386. Term apomixis was coined by ....
387. Apomix was discovered by which scientist ....
388. Formation of multiple embryos in a single seed is called ....
389. Optimum temperature for callus formation for grafting ....
390. Most common and effective rooting growth regulator for cutting ....
391. What is the main role of interstock ....

392. The best planting material in strawberry is ....
393. Phalsa is comercially propageted by ....
394. Pineapple is commercially propagated by ....
395. Gladiolus propagated by ....
396. Commerially propagetion of Bougainvillea ....
397. Papaya is commercially propagated by ....
398. Hard wood cutting can be used to propagated fruit crop ....
399. Grape is commercially propageted by ....
400. Guava is commercial propageted by ....
401. Phalsa is commercial propageted by ....
402. Cerries is commercially propageted through ....
403. National flower of india is ....
404. Queen of flower is ....
405. Dahlia is propagated by ....
406. Ber is commercially propagated by ....
407. Karonda is commercially propagated by ....
408. Dwarfing rootstock of apple is ....
409. Standardized rootstock of guava is ....
410. Semi hard wood cutting can be used to .... fruit crope.
411. Asexually propagated plants are true to type to their....................
412. Disadvantages of asexually propagation methods plants have ....
413. The length of hard wood cutting is kept to ....
414. The lenghth of semi hard wood cutting is kept to ....
415. Semi hard wood cutting treatment of IBA @ of ....
416. Stem cutting is sucsesfull in propagation of ....
417. Ring budding is ....
418. Epicotyle grafting is also known as ....
419. Soft wood cutting is very sucsesfull technique of .... grafting
420. Inarching is called as .... of grafting
421. Ber fruits crop is a time budding ....
422. The best planting meterial in strawberry is ....
423. Fig is commercially propagated by....................

424. Aonla propagation through budding is time of moth ....

425. Commercially propagation of apple is ....

426. Gladeolus is propagated by ....

427. Dahlia is propagated by ....

428. Phalsa is commercially propagated by ....

429. Drarfing rootstock of apple is ....

430. Semi hard wood cutting can be used to .... fruit crop

2. Multiple choice:

1. **Export variety of mango is**

   (a) Langra (b) Dashehri

   (c) Banganpalli (d) Alphonso

2. **In India, the leading state in area and production of mango**

   (a) Bihar (b) Uttar Pradesh

   (c) West Bengal (d) Andhra Pradesh

3. **Time of flower bud differentiation in mango**

   (a) Jan-Feb (b) April- May

   (c) June- July (d) Oct.-Nov

4. **Mosambi is successfully cultivated in**

   (a) Haryana (b) Punjab

   (c) Assam (d) Maharashtra

5. **Sweet orange is fruit with**

   (a) Tight Skinned (b) Loose skinned

   (c) Partially loose (d) Partially tight

6. **Mango hybrid free from spongy tissue problem**

   (a) ArkaAruna (b) Ratna

   (c) ArkaPuneet (d) ArkaAnmol

7. **Color development in sweet orange favored by temperature-**

   (a) High in winter (b) Low in winter

   (c) Mild summer (d) Hot and humid summer

8. **Mosambi cultivation is preferred in**

   (a) Maharashtra (b) Rajasthan

   (c) Punjab (d) Himachal Pradesh

9. Sweet oranges are generally consumed as

(a) Table fruit
(b) Squash
(c) Pickle
(d) Juice

10. Common rootstock for budding of mandarin

(a) C. jambhri
(b) C. sinensis
(c) Citrus limon
(d) Rangpur lime

11. Sweet orange prefers soil-

(a) Loam
(b) Mediam loan
(c) Clay
(d) Clay loam

12. Commercial method of propagation of Aonla is –

(a) Veneer grafting
(c) Patch budding
(c) T-budding
(d) Air layering

13. Which is the largest Aonla producing district in the Uttar Pradesh.

(a) Faizabad
(b) Varanasi
(b) Pratapgarh
(d) Azamgarh

14. Which district is the largest producer and processing of Aonla in the Uttar Pradesh

(a) Faizabad
(b) Chandauli
(c) Pratapgarh
(d) Varanasi

15. Which type of fruit Bael is –

(a) Drupe
(b) Berry
(c) Stone
(d) Pome

16. In India, which of the following fruits has highest production –

(a) Mango
(b) Citrus
(c) Aonla
(d) Bael

17. NB-7 is cultivar of a–

(a) Grape
(b) Mango
(c) Bael
(d) Citrus

18. Fruiting time of Bael is –

(a) Jan-Feb
(b) Oct.-Nov
(c) April-May
(d) None of these

19. Cherimoya is related to -

(a) Wood apple
(b) Rose apple
(c) Custard apple
(d) Apple

20. Flying Dragon, a dwarfing rootstock of citrus, is susceptible to-

(a) Citrus tristeza (b) Phytophthora root rot

(c) Citrus nematode (d) Iron chlorosis

21. "Bangkok Golden Apple" is a variety of-

(a) Guava (b) Pineapple

(c) Avocado (d) Pomegranate

22. 'Chicken-tongue' seed formation in litchi is due to-

(a) Parthenogenesis (b) Stenospermocarpy

(c) Failure of pollination (d) Nutritional deficiency

23. The somatic chromosome number (2n) of Illaichi cultivar of ber (Zizyphusmauritiana) is-

(a) 12 (b) 24

(c) 48 (d) 96

24. Which of the following sapota cultivar can produce both round and oval shaped fruits on the same tree?

(a) Cricket Ball (b) Kalipatti

(c) Oval (d) Bhuripatti

25. Fruit cracking is related to which fruit-

(a) Guava (b) Bael

(c) Mango (d) Sapota

26. The main photosynthetic product in apple is-

(a) Sorbitol (b) Mannitol

(c) Fructose (d) Sucrose

27. Seeded variety of Grape is-

(a) Anab-e-shahi (b) Beauty seedless

(c) Thomson seedless (d) Pusalalima

28. Black tip is related to-

(a) Aonla (b) Bael

(c) Mango (d) Guava

29. Polyembrionic variety of Mango-

(a) L-49 (b) Chakaiya

(c) Langra (d) Allahabadisafeda

30. Citrus reticulata is-

(a) Citron (b) Lime

(c) Mandarin (d) Lemon

31. Annona belongs to family-

(a) Sapotaceae (b) Rosaceae
(c) Annonaceae (d) Myrtaceae

32. Botanical name of Grape-

(a) Vitisvinifera (b) Averrhoabilimbi
(c) Annonasquamosa (d) Passifloraedulis

33. Chromosome no. of Banana is-

(a) 24 (b) 56 (c) 26 (d) 22

34. Litchi is ________ type of fruit-

(a) Drupe (b) Nut (c) Berry (d) Sorosis

35. Origin place of Jack fruit is-

(a) North America (b) India (c) China (d) Australia

36. Edible part of Litchi is-

(a) Mesocarp (b) Aril (c) Seeds (d) Rind

37. Plant which are used as wind break.

(a) Desi mango (b) Guava (c) Ber (d) Karonda

38. Commercial propagation is bael fruit crop.

(a) Root cutting (b) Patch budding
(c) Grafting (d) Seedling

39. Regular pruning is needed in.

(a) Apple (b) Guava (c) Peach (d) Mango

40. Suitable time for the propagation of apple is

(a) Feb – March (b) Dec – Jan
(c) July – Aug (d) April – May

41. High density planting of fruits plant is useful because of

(a) High productivity (b) Use of inputs
(c) Easy harvest (d) Distance

42. Mango is commercially propagation technique

(a) Veneer grafting (b) Inarching
(c) Epicotyle (d) Seeds

43. Jack fruit is commercially propagation tecknique

(a) Seed (b) Patch budding
(c) Clapt grafting (d) Cutting

44. Commercial type of propagating structures in nursery

(a) Glass house (b) Green house

(c) Lath house (d) Bottom heat

45. Low cost technique structure in nursery grow

(a) Net house (b) Plastic house

(c) Low tennel (d) Glass house

46. Gooseberry is commercial propagation technique

(a) Seedling (b) Runner

(c) Cutting (d) None of these

47. Grape is propagated through

(a) Hard wood cutting (b) Soft wood cutting

(c) Root cutting (d) None of these

48. Sweet orange is propagated through

(a) T budding (b) H budding

(c) Air layering (d) None of these

49. Which one of the following gas is called as ripening hormons

(a) CO_2 (b) O_2 (c) C_2H_4 (d) H_2

50. Litchi is commercially propagation technique

(a) Simple layering (b) Serpentin layering

(c) Mount layering (d) Air layering

51. Aonla is commercially propagation technique

(a) Ring budding (b) T budding

(c) Patch budding (d) H budding

52. Karonda is commercially propagated by

(a) Cutting (b) Seed (c) Layering (d) Budding

53. Coffee produced obtained from

(a) Leaves (b) Stem (c) Flower (d) Seeds

54. Arecanunt is the largest producing state

(a) J&K (b) Karnataka (c) Tamilnadu (d) Kerela

55. Botanical name of Mangosteen

(a) Garciniacambogia (b) Chestuamollissima

(c) Garciniamangostana (d) None of these

56. Botanical name of Bread fruit

(a) Perseaamericana
(b) Malphigiapunicifolia
(c) Artocarpusaltilis
(d) None of these

57. Cape gooseberry is propagated by

(a) Cutting (b) Layering (c) Seeds (d) Grafting

58. Duo dichogamy is observed in

(a) Pistachio nut (b) Walnut (c) Chestnut (d) Coconut

59. Type of incompability found in almond is

(a) Sporophytic (b) Gametophytic (c) Both

60. Type of apomixis present in citrus

(a) Non recurrent
(b) Recurrent
(c) Parthenocarpy
(d) All of above

61. Nugget is dwarf cultivar of ....

(a) Apple (b) Pear (c) Peach (d) Plum

62. Rangpurlime and acid lime is propagated by

(a) Seeds
(b) Layering
(c) Hardwood cutting
(d) Grafting

63. Most commonly propagation method in kiwifruit

(a) Seeds (b) Layering (c) Cuttings (d) Stooling

64. Endosperm like tissue in seed developing from nucellus is.......

(a) Endosperm (b) Cotyledon (c) Perisperm (d) Mericarp

65. Dormancy due to factor within embryo

(a) Endogenous
(b) Exogenous
(c) Combination
(d) None of above

66. During stratification most important factors is/are

(a) Temperature
(b) Duration
(c) Media
(d) All of above

67. Stage at which seed achives its maximum dry weight and has maximum germination

(a) Physiological maturity
(b) Edible maturity
(c) Harvest maturity
(d) None of above

68. Seed vigor is affected by

(a) Time of seed stored
(b) Type of seed stored
(c) Storage environment
(d) All of above

69. Tetrazolium test is to check seed viability

(a) Chemical test (b) Enzyme test

(c) Germination test (d) a & c

70. Which fig is able to produce pollen

(a) san pedro (b) smyrna (c) capri (d) dauldabad

71. Girdling is most common in

(a) litchi (b) mango (c) grape (d) citrus

72. Indian jujube is propagated by

(a) Veneer grafting (b) T- budding (c) Seed (d) None

73. Paja commonly used rootstock for

(a) Cherry (b) Apricot (c) Almond (d) Peach

74. `Dwarfing rootstock used in HDP of ber

(a) Z. jujuba (b) Z. nummularia

(c) Z. lotus (d) Z. rotundifolia

75. Loquat is propagated by

(a) Seed (b) Patch budding

(c) Sucker (d) T – budding

76. In grafting the rooted plant is called

(a) Root stock (b) Transplant

(c) Scion (d) None of above

77. In grafting stem cutting from donor plant is called

(a) Transplant (b) Scion (c) Budgraft (d) Rootstock

78. Which of following method is suitable for combining the desirable characters of two plants together in a single plant

(a) Air layering (b) Cutting (c) Grafting (d) All of these

79. Compound layering also called as

(a) Air layering (b) Serpentine layer

(c) Trench layering (d) Mound layering

80. Air layering is also called as

(a) Gootee (b) Chinese layering

(c) Pot layering (d) All of these

81. Rooting in stem cuttings are stimulated by using

(a) Jasmonic acid (b) ABA (c) Ethylene (d) IAA

82. Best time of budding in ber

(a) Feb- March (b) July -Aug (c) May-June (d) Sept-Oct

83. Best time for propagation of apple

(a) Dec-Jan (b) July –Aug (c) Feb- March (d) April-May

84. Which of following is the commercial method of propagation of phalsa

(a) Seed (b) Cutting (c) Budding (d) Layering

85. Sapota is commercially propagated by

(a) Inarching (b) Seed (c) Air layering (d) Cutting

86. Passion fruit is commercially propagated by

(a) Seed (b) Cutting (c) Budding (d) Layering

87. Semi hard wood cutting should be

(a) One year old twigs (b) One season old twigs

(c) New growth (d) All of above

88. Semi hardwood cutting is

(a) Fig (b) Pear (c) Rose (d) All of above

89. Length of soft wood cuttings should be

(a) 5-10cm (b) 10-15 cm (c) 15-25cm (d) 25-30cm

90. Very vigorous rootstock of apple

(a) M-12 (b) M-13 (c) M-16 (d) All of these

91. Standard rootstock of apple against wooly aphid is

(a) Malling - Merton (b) DX-1

(c) PT-1 (d) None of these

92. Peach is mainly propagated by

(a) Shield budding (b) Air layering

(c) Cutting (d) None of these

93. Which of following rootstock of grape is useful under high salt content and nematodes

(a) Rumani (b) Olour

(c) Dogridge (d) None of above

94. Which of following suckers are used in propagation of banana

(a) Tall suckers (b) Dwarf suckers

(c) Sword suckers (d) Water suckers

95. Rootstock used in propagation of ber

(a) Jherberi (b) M-8 (c) X-7 (d) All of these

96. Root stock of grape resistant to phylloxera

(a) St. George (b) RipariaGloire

(c) K5BB (d) All of these

97. Stooling is also called as

(a) Sexual propagation (b) Cutting

(c) Gooti layering (d) Mound layering

98. Tap roots of plants are also called as

(a) Secondry roots (b) Adventitious roots

(c) Primary roots (d) Tertiary roots

99. Part of stem from which leaf arises is called

(a) Node (b) Internode (c) Apex (d) Bud

100. Bud which develops into axis of a leaf is

(a) Auxillary buds (b) Terminal buds

(c) Adventitious buds (d) Naked buds

101. Dwarfing rootstock for apple in heavy soils

(a) M-7 (b) M-9 (c) M-11 (d) Crab apple

102. Following species of citrus can be propagated through cutting

(a) Sweet orange (b) Sweet lime

(c) Kaghzi lime (d) None of above

103. Following species of citrus can be propagated through layering

(a) Sweet orange (b) Sweet lime

(c) Kaghzi lime (d) None of above

104. Rubber plant can be propagated through

(a) Aerial layering (b) Cutting

(c) All of above \ (d) None of above

105. silver oak produces seed after age of

(a) 1 year (b) 5 year (c) 10 year (d) 1 year

106. Name the method of asexual propagation in plants in which callus is formed

(a) Regeneration (b) Micro propagation

(c) Fragmentation (d) Vegetative propagation

107. In which artificial propagation method stock and scion are involved

(a) Tissue culture (b) Cuttings (c) Grafting (d) Layering

108. Root cuttings are practised in fruit crops

(a) Black berry (b) Raspberry

(c) Both a & b (d) None of above

109. Hardwood cuttings are used in

(a) Fig (b) Olive

(c) Grapes (d) All of above

110. Top budding is practised in

(a) Cahewnut (b) Mango

(c) Both a & b (d) None of above

111. Tip layering is commonly practised in

(a) Blackberry (b) Raspberry (c) Dew berry (d) All of these

112. Polyembrony was reported in which fruit crops

(a) Citrus (b) Mango (c) Jamun (d) All of these

113. Nucellarpolyembryony is found in

(a) Citrus (b) Litchi (c) Pomegranate (d) Coconut

114. Rapid multiplication of mango can be done by which type of grafting

(a) Wedge grafling (b) Veneer grafting

(c) Epicotyl grafting (d) Inarching

115. Cleft grafting is commonly employed for rejuvenating temperate orchards by top working in which fruit crops

(a) Walnut (b) Pecan nut

(c) Hazle nut (d) All of above

116. Fruit crop which develop fruit parthenocarpically still produce viable seeds

(a) Mangosteen (b) Strawberry

(c) Both (d) None of above

117. The best method of irrigation for mango planted in sandy soil is

(a) Sprinkler (b) Drip (c) Channel (d) Bed

118. In hexagonal planting system how much more plants can accommodate then the squire system.

(a) 10 % (b) 15% (c) 25% (d) 40%

119. M - 9 Dwarfing rootstock of apple has originated as a result of

(a) Cross of M16 X M27 (b) Cross of 27 x Northern spy.

(c) Cross of northern spy x M16 (d) Chance seedling selection

120. Plant which are used as wind break.

(a) Desi mango (b) Guava (c) Ber (d) Karonda

121. Commercial propagation is bael fruit crop.

(a) Root cutting (b) Patch budding (c) Grafting (d) Seedling

122. Regular pruning is needed in.

(a) Apple (b) Guava (c) Peach (d) Mango

123. Suitable time for the propagation of apple is

(a) Feb – March (b) Dec – Jan

(c) July – Aug (d) April – May

124. High density planting of fruits plant is useful because of

(a) High productivity (b) Use of inputs

(c) Easy harvest (d) Distance

125. Mango is commercially propagation technique

(a) Veneer grafting (b) Inarching (c) Epicotyle (d) Seeds

126. Jack fruit is commercially propagation tecknique

(a) Seed (b) Patch budding

(c) Clapt grafting (d) Cutting

127. Commercial type of propagating structures in nursery

(a) Glass house (b) Green house

(c) Lath house (d) Bottom heat

128. Low cost technique structure in nursery grow

(a) Net house (b) Plastic house

(c) Low tennel (d) Glass house

129. Gooseberry is commercial propagation technique

(a) Seedling (b) Runner

(c) Cutting (d) None of these

130. Grape is propagated through

(a) Hard wood cutting (b) Soft wood cutting

(c) Root cutting (d) None of these

131. Sweet orange is propagated through

(a) T budding (b) H budding

(c) Air layering (d) None of these

132. Which one of the following gas is called as ripening hormons

(a) CO_2 (b) O_2 (c) C_2H_4 (d) H_2

133. Litchi is commercially propagation technique

(a) Simple layering (b) Serpentin layering

(c) Mount layering (d) Air layering

134. Aonla is commercially propagation technique

(a) Ring budding (b) T budding

(c) Patch budding (d) H budding

135. Karonda is commercially propagated by

(a) Cutting (b) Seed (c) Layering (d) Budding

136. Coffee produced obtained from

(a) Leaves (b) Stem (c) Flower (d) Seeds

137. Arecanunt is the largest producing state

(a) J&K (b) Karnataka (c) Tamilnadu (d) Kerela

138. Botanical name of Mangosteen

(a) Garcinia cambogia (b) Chestua mollissima

(c) Garcinia mangostana (d) None of these

139. Botanical name of Bread fruit

(a) Persea americana (b) Malphigia punicifolia

(c) Artocarpus altilis (d) None of these

140. Mango is commercially propagation technique

(a) Veneer grafting (b) Inarching (c) Epicotyle (d) Seeds

141. Jack fruit is commercially propagation tecknique

(a) Seed (b) Patch budding

(c) Clapt grafting (d) Cutting

142. Commercial type of propagating structures in nursery

(a) Glass house (b) Green house

(c) Lath house (d) Bottom heat

143. Low cost technique structure in nursery grow

(a) Net house (b) Plastic house

(c) Low tennel (d) Glass house

144. Gooseberry is commercial propagation technique

(a) Seedling (b) Runner

(c) Cutting (d) None of these

145. Litchi is commercially propagation technique

(a) Simple layering (b) Serpentin layering

(c) Mount layering (d) Air layering

146. Aonla is commercially propagation technique

(a) Ring budding (b) T budding

(c) Patch budding (d) H budding

147. Karonda is commercially propagated by

(a) Cutting (b) Seed (c) Layering (d) Budding

148. Plant which are used as wind break.

(a) Desi mango (b) Guava (c) Ber (d) Karonda

149. Commercial propagation is bael fruit crop.

(a) Root cutting (b) Patch budding (c) Grafting (d) Seedlling

150. Regular pruning is needed in.

(a) Apple (b) Guava (c) Peach (d) Mango

151. Grape is propagated through

(a) Hard wood cutting (b) Soft wood cutting

(c) Root cutting (d) None of these

152. Sweet orange is propagated through

(a) T budding (b) H budding

(c) Air layering (d) None of these

153. Which one of the following gas is called as ripening hormons

(a) CO_2 (b) O_2 (c) C_2H_4 (d) H_2

154. Suitable time for the propagation of apple is

(a) Feb – March (b) Dec – Jan

(c) July – Aug (d) April – May

155. High density planting of fruits plant is useful because of

(a) High productivity (b) Use of inputs
(c) Easy harvest (d) Distance

156. Coffee produced obtained from

(a) Leaves (b) Stem (c) Flower (d) Seeds

157. Utter Pradesh fruit Nursery act,

(a) 1776 (b) 1974 (c) 1975 (d) 1969

158. Date palm is commercially propagation metho

(a) Off shoot (b) Grafting
(c) Bridge Grafting (d) None of these

159. Banana propagation by

(a) Suckers (b) Rhizoms
(c) Slips (d) None of these

3. Write 'True' or 'False'

1. Yellow color in Bael is due to Anthocyanin ( )
2. Bael is known as king of Arid fruits ( )
3. Patch budding is commercial propagation method of Annona ( )
4. Aonla is propagated by seeds ( )
5. Bael is propagated by seeds ( )
6. Planting Distance of Bael is 1x1 meter ( )
7. Jolo is variety of Mangosteen ( )
8. Pruning of cane type grape variety should be done leaving 6-8 buds. ( )
9. Aonlafruit belongs to familyAnacardiaceae. ( )
10. Drupe is the type of fruit of Mango. ( )
11. Family of Rose apple is Rutaceae. ( )
12. Beauty seedless is a variety of Grape. ( )
13. Botanical name of Carambola is Averrhoacarambola. ( )
14. Chromosome no. of Aonla is 24. ( )
15. Botanical name of Durian is Duriozibethinus. ( )
16. Mango is commercially propagated by Inarching. ( )
17. Edible part of Citrus is Placental hairs. ( )
18. Sapotabelong to family Sapotaceae. ( )

19. Time of flower bud initiation in mango is Oct.- Nov. month ( )
20. Thompson Seedless is ideal cv. for raisin making ( )
21. Amrapali is cross between Neelam x Dashehari ( )
22. Chausa is a alternate bearing variety of mango ( )
23. Ratna is a hybrid developed at IIHR Bangaluru ( )
24. The inflorescence of banana is Catkin) ( )
25. Mandarin is tight skinned orange ( )
26. Planting distance of Sweet orange is 4x4 m ( )
27. Eureka is variety of Kagzi lime ( )
28. Citrus paradisi is botanical name of grape fruit ( )
29. Cane prune grape variety bears fruit on distal 10-12 bud ( )
30. Aonla is known as Indian gooseberry ( )
31. Origin center of Carambola is South Asia ( )
32. Kagzi is variety of Bael. ( )
33. Bael belong to family Myrtaceae. ( )
34. Origin place of Corombola is India. ( )
35. Fig is gynodioecious spp. ( )
36. Corombola is rich source of oxalic acid. ( )
37. Fruit of the 21st century is called Aonla. ( )
38. Marmelosin is present in Durian. ( )
39. Golden star (Hawaii) is popular variety of Corombola. ( )
40. Strawberry basic chromosome number (X=7). ( )
41. Mangosteen harvesting time is between Nov-Dec. ( )
42. Mangosteen commercially propagated by seeds. ( )
43. Head quarter of AICRP of Arid Zone Fruit is in Bangalore, Karnataka. ( )
44. Head quarter of AICRP of Sub-tropical Fruits is in Lucknow, Uttar Pradesh. ( )
45. Central Institute of Horticulture situated in Mediziphema, Nagaland. ( )
46. Highest fibre content is in Fig. ( )
47. Annona is seedless due to ovule sterility. ( )
48. Yellow colour of seed is due to Xanthophylls. ( )
49. Apple is non climateric fruit. ( )

50. Apple scab is caused by Bacteria. ( )
51. Origin of Peach is Afghanistan. ( )
52. Mealy bug is major insect of mango. ( )
53. ArkaneelamandAlphonso are poly embryonic variety of mango. ( )
54. Use of Bordeux mixture is effective method to control mango mealy bug. ( )
55. Ripe fruit ofChausavariety of mango remain green in colour. ( )
56. Fruit cracking is serious problem in Bael. ( )
57. Name of Mango is because of its sweet flavour. ( )
58. Canker is serious problem of Sapota. ( )
59. Botanical name of Rambutan is Embelicaofficinalis. ( )
60. Chromosome no. of Bael is 24. ( )
61. Family of Durian is Oxilidaceae. ( )
62. Aonla is commercially propagated by grafting. ( )
63. Edible part of Bael is seeds. ( )
64. Durian belong to family Annonaceae. ( )
65. Family of Mangosteen isRosaceae. ( )
66. Family of Avocado is Lauraceae. ( )
67. Sapota Chromosome number (2n) = 56. ( )
68. Black tip is a physiological disorder of Mango. ( )
69. Type of fruit in pomegranate is berry. ( )
70. Botanical name of Sapota is Manilkarazapota. ( )
71. Breadfruit belongs to family Rosaceae. ( )
72. Chromosome no. of Bilimbi is 18. ( )
73. Loquat is a stone type of fruit. ( )
74. Type of fruit in Durian issorosis. ( )
75. Mango fruit is commonly known as king of fruits. ( )
76. Chromosome no. of Citrus is 18. ( )
77. Pusalalimais a variety of Mango. ( )
78. Rambutan is commercially propagated by grafting. ( )
79. Origin place of Rambutan is Malaysia. ( )
80. Bael is commercially propagated by seeds. ( )
81. Edible part of Aonla is mesocarp. ( )

82. Type of inflorescence in Mango is panicle. ( )
83. Botanical name of Aonla is Embelicaofficinalis. ( )
84. Edible part of Annona is epicarp. ( )
85. Predominant acid present in Peach ismallic acid. ( )
86. Flower color of Pear is white. ( )
87. Bael fruit is plant symbol of Lord Shiva. ( )
88. Type of fruit in Fig is berry. ( )
89. Predominant pollination in Pomegranate done by honey bee. ( )
90. Strawberry Chromosome number (2n) = 28. ( )
91. Capsacinis responsible for flavor of Strawberry. ( )
92. Type of inflorescence in Aonla is panicle. ( )
93. Botanical name of Rose apple isSyzygiumjambos. ( )
94. Rose apple belongs to family Rosaceae. ( )
95. Chromosome no. of Avocado is 24. ( )
96. Mango is a drupe type of fruit. ( )
97. Type of fruit in Citrus is Hesperidium. ( )
98. Cracking is major disease of Citrus. ( )
99. Chakaiyaand L-49 are varieties of Guava. ( )
100. Use of bordeux mixture is effective method to control Citrus canker. ( )
101. Mosambicultivar of Citrusis used for juice making at maturity stage. ( )
102. Tristeza is serious problem in Citrus. ( )
103. Leaves ofBael fruit is used as offerings to lord Shiva. ( )
104. Malformation is serious problem of Citrus. ( )
105. Mandarin is propagated commercially by cutting. ( )
106. Anab-e-shahi is spur pruned variety of grape. ( )
107. Poovancultivarof banana is most suitable for table purpose. ( )
108. Chromosome number of Mangiferaindica is 56. ( )
109. Family of Litchi is Rosaceae. ( )
110. Family of Passion fruit is Passifloraceae. ( )
111. Write any two varieties of GrapePerlett and Beauty seedless. ( )
112. Lemon is propagated commercially by seeds. ( )
113. Thompson seedless is spur pruned variety of grape. ( )

114. Beauty seedless is most suitable variety for raisin making. ( )
115. Chromosome number of Vitisvinifera is 26. ( )
116. Pruning of spur type grape variety should be done leaving 3-4 buds. ( )
117. Annonafruit belongs to familyAnacardiaceae. ( )
118. Hesperidium is the type of fruit of Citrus. ( )
119. Family of Mangosteen is Rutaceae. ( )
120. NB-9 is a variety of Bael. ( )
121. Litchi is commercially propagated by Layering. ( )
122. Origin place of Grape is Mediterranean region. ( )
123. Bael is commercially propagated by seeds. ( )
124. Edible part of Litchi is aril. ( )
125. Type of fruit in Grape is drupe. ( )
126. Botanical name of Bilimbi is Anacardiumbilimbi. ( )
127. Edible part of Papaya is mesocarp. ( )
128. Predominant acid present in Citrus is tartaric acid. ( )
129. Flower color of Bael is purple. ( )
130. Mangosteen fruit is known as queen of fruits. ( )
131. Type of fruit in Guava is Balausta. ( )
132. Pineappleis Commercial propagated byslips and suckers. ( )
133. Central insttitue of arid horticulture is located at Bikaner ( )
134. The National fruit of india is Mango ( )
135. Seed of graps fruit are smooth and polyembryonic ( )
136. Ber is heavily pruned fruit crop ( )
137. Cabbage is propageted heads ( )
138. The plants produced from vegetative means are tree to type ( )
139. Date palm is propagated by off shoot ( )
140. Pine apple is propagated by runner ( )
141. Litchi is propagated by air layering ( )
142. NAA is fruit thining growth regulator ( )
143. Best time of planting of orcherd is july – august ( )
144. High density orcherd have more water requirment ( )
145. Commercially propagation method of aonla is ring budding ( )

146. Which of the following chemical is used for termite control is cloripyeriphas ( )
147. Polyembryony is found in mango ( )
148. Old mango orcherd are rejuvenated by top grafting ( )
149. The rootstock in thepropation of ber is jerberi ( )
150. The dwarfing rootstock for mango is olour ( )
151. China is the largest producer of fruits in the world ( )
152. Best time for the budding in ber is Dec – Jan ( )
153. Central insttitue of arid horticulture is located at Bikaner ( )
154. The National fruit of india is Mango ( )
155. Seed of graps fruit are smooth and polyembryonic ( )
156. Ber is heavily pruned fruit crop (T/F)
157. Cabbage is propageted heads (T/F)
158. The plants produced from vegetative means are tree to type (T/F)
159. Date palm is propagated by off shoo (T/F)
160. Pine apple is propagated by runner (T/F)
161. Litchi is propagated by air layering (T/F)
162. NAA is fruit thining growth regulator (T/F)
163. Best time of planting of orcherd is july – august (T/F)
164. High density orcherd have more water requirment (T/F)
165. Commercially propagation method of aonla is ring budding (T/F)
166. Which of the following chemical is used for termite control is cloripyeriphas (T/F)
167. Polyembryony is found in mango (T/F)
168. Old mango orcherd are rejuvenated by top grafting (T/F)
169. The rootstock in the propation of ber is jerberi (T/F)
170. The dwarfing rootstock for mango is olour (T/F)
171. China is the largest pr oducer of fruits in the world (T/F)
172. Best time for the budding in ber is Dec – Jan (T/F)
173. CIH central institue of horticulture Mediziphema Nagaland (T/F)
174. NHM National Horticulture Mission 2005 – 06 (T/F)
175. Seed of grapes are smooth and pollyembryonic (T/F)

176. Ber is heavy pruned fruit crop (T/F)
177. Date palm is propagated by off shoot (T/F)
178. NAA is fruit thining grouth regulator (T/F)
179. Best time of planting of orcherd is july – aug (T/F)
180. High density orchard have more water requirement (T/F)
181. Polyembryony is found in mango (T/F)
182. Polyembryony is rootstock of mango – Bellary (T/F)
183. Caging techniques developed by Sharma et all. (T/F)
184. Amraali cross betwin Neelam x Dashehri (T/F)
185. Off season mango fruit maturity in Kanya kumari district of Tamilnadu in Jan. – fab. (T/F)
186. Epicotyle grafting in mango 24 – 27 days seedling (T/F)
187. Litchi is propagated by air layering (T/F)
188. National fruit of Afganistan is Jaickfruit (T/F)
189. Jaick fruit is commercially propagated by Shield budding (T/F)
190. Wood apple is commercially propagated by budding (T/F)
191. Lemon is commercially propagated by budding (T/F)
192. Raised bed up to hieght of above the ground level 15cm (T/F)

4. Matching to the correct answer:-

A	B
A. Bael	1. Sucker or silips
B. Pineapple	2. Air layering
C. Lemon	3. Patch budding
D. Strawberry	4. Stooling
E. Guava	5. Stolene
F. Ber	6. Hard wood cutting
G. Grape	7. Off shoot
H. Mango	8. Ring budding
I. Date palm	9. Veneer grafting
J. Root harmons	10. Nemaguard
K. Ripening hormons	11. Dogridge
L. Peach rootstock	12. IBA

M. Pear roostock
N. Grape rootstock
O. Guava rootstock
P. Mango rootstock
Q. Apple rootstock
R. Aonla rootstock
S. Temple rootstock
T. Custerd apple

13. Ethyline
14. Tomato
15. Lucknow 49
16. Mango
17. Rangpur lime
18. Grape
19. Kurrukan
20. Annonaglabra

B. Methods	Plants
I. Soft wood cutting	(A) pineapple
II. Semi hard wood cutting	(B) banana
III. Hard wood cutting	(C) fig
IV. Air layering	(D) cherry
V. Stool layering	(E) Lemon
VI. Rooting hormones	(F) Litchi
VII. Pot layering	(G) IBA
VIII. Marcottage	(H) Guava
IX. Rhizome	(I) loquat
X. Sucker	(J) Gootee

C. Matching-

1. Bael	1. Sucker or silips
2. Pineapple	2. Air layering
3. Lemon	3. Patch budding
4. Strawberry	4. Stooling
5. Guava	5. Stolene
6. Ber	6. Hard wood cutting
7. Grape	7. Off shoot
8. Mango	8. Ring budding
9. Date palm	9. Veneer grafting
10. Root harmons	10. Nemaguard
11. Ripening hormons	11. Dogridge
12. Peach rootstock	12. IBA

13. Pear roostock
14. Grape rootstock
15. Guava rootstock
16. Mango rootstock
17. Apple rootstock
18. Aonla rootstock
19. Temple rootstock
20. Custerd apple
21. 61. Guava rootstock
22. 62. Mango rootstock
23. 63. Apple rootstock
24. 64. Tapka
25. 65. Temple rootstock
26. 66.Truss
27. 67. Bael
28. 68. Pineapple
29. 69. Lemon
30. 70. Strawberry
31. 71. Guava
32. 72. Ber
33. 73. Grape
34. 74. Mango
35. 75. Date palm
36. 76. Root harmons
37. 77. Ripening hormons
38. 78. Peach rootstock
39. 79. Pear roostock
40. 80. Grape rootstock

13. Ethyline
14. Tomato
15. Lucknow 49
16. Mango
17. Rangpur lime
18. Grape
19. Kurrukan
20. Annona glabra
21. Lucknow 49
22. Mango
23. Rangpur lime
24. Grape
25. Kurrukan
26. Quince
27. Sucker or silips
28. Air layering
29. Patch budding
30. Stooling
31. Stolene
32. Hard wood cutting
33. Off shoot
34. Ring budding
35. Veneer grafting
36. Nemaguard
37. Dogridge
38. IBA
39. Ethyline
40. Tomato

❐❐❐

9 789354 614750

Printed at Repro India Ltd.